CW00417714

WILDLIFE IN THE
_____

WILDLIFE
in the
Anthropocene

Conservation after Nature

JAMIE LORIMER

University of Minnesota Press
Minneapolis · London

An earlier version of chapter 1 was published as "Multinatural Geographies for the An- thropocene," *Progress in Human Geography* 36, no. 5 (2012): 593–612. An earlier version of chapter 2 was published as "Nonhuman Charisma," *Environment and Planning D: Society and Space* 25, no. 5 (2007): 911–32. An earlier version of chapter 3 was published as "What About the Nematodes? Taxonomic Partialities in the Scope of UK Biodiversity Conservation," *Social and Cultural Geography* 7, no. 4 (2006): 539–58. An earlier version of chapter 4 was published as "Counting Corncrakes: The Affective Science of the UK Corncrake Census," *Social Studies of Science* 38, no. 3 (2008): 377–405. An earlier version of chapter 5 was coauthored with Clemens Driessen and published as "Wild Experiments at the Oostvardersplassen: Rethinking Environmentalism for the Anthropocene," *Trans- actions of the Institute of British Geographers* 39, no. 2 (2014): 169–81. An earlier version of chapter 6 was published as "Moving Image Methodologies for More-Than-Human Geographies," *Cultural Geographies* 17, no. 2 (2010): 237–58. Portions of chapter 7 were previously published as "International Conservation Volunteering and the Geographies of Global Environmental Citizenship," *Political Geography* 29, no. 6 (2010): 311–22, and "International Conservation Volunteering: What Difference Does It Make?" *Oryx* 43, no. 3 (2009): 352–60. Portions of chapter 8 were previously published as "Living Roofs and Brownfield Wildlife: Towards a Fluid Biogeography of UK Nature Conservation," *Environment and Planning A* 40, no. 9 (2008): 2042–60.

Copyright 2015 by the Regents of the University of Minnesota

Published by the University of Minnesota Press
111 Third Avenue South, Suite 290
Minneapolis, MN 55401-2520
http://www.upress.umn.edu

Library of Congress Cataloging-in-Publication Data

Lorimer, Jamie.
    Wildlife in the Anthropocene : conservation after nature / Jamie Lorimer.
    Includes bibliographical references and index.
    ISBN 978-0-8166-8107-5 (hc : alk. paper)
    ISBN 978-0-8166-8108-2 (pb : alk. paper)
1. Nature conservation. 2. Nature—Effect of human beings on. 3. Animals—Effect of human beings on. 4. Ecology. I. Title.
    QH75.L62   2015
    333.72—dc23

                                                                            2014019919

Printed in the United States of America on acid-free paper

The University of Minnesota is an equal-opportunity educator and employer.

21 20 19 18 17 16 15   10 9 8 7 6 5 4 3 2 1

To my parents and siblings for their love, for teaching me to think independently, and for their respect, even when they don't understand what it is that I say and do. To Magali, to Amelie, and to Louis for showing me that a wild life need not be out there, singular, predictable, or all that reasonable.

CONTENTS

INTRODUCTION

After the Anthropocene

Geologists argue that our planet has entered the Anthropocene.[1] A new epoch began once humans became an earth-changing force, capable of leaving their signature in the fossil record.[2] There is a growing acceptance of this term among scientists, politicians, and other elites, which accompanies a recognition that there are few places, forms, and processes on this planet that do not bear the traces of human activity.[3] This is not, however, the triumph of Enlightenment science. Nature has not finally been known, tamed, and rationally ordered. Instead, the unforeseen, deleterious, and unequal consequences of these planetary activities are an established source of concern.

This diagnosis of the Anthropocene is revolutionary, akin to the shocking thoughts of Copernicus, Lyell, and Darwin.[4] Many cultures are still coming to terms with an understanding of the world as ancient, one of many and not built for us. Evolution continues to prove challenging to familiar figures of the created or at least uniquely social human. Now, we are being depicted as geological actors, entangled within and responsible for a powerful, unstable, and unpredictable planetary system. Unsurprisingly perhaps, for some publics the magnitude and consequences of our geological entanglements are proving hard to accept.[5]

The possibility of human planetary impacts provokes less cognitive dissonance among conservationists—who are the focus of this book. Such impacts have been staple concerns since at least the nineteenth century. But the diagnosis of the Anthropocene challenges the modern figure of Nature that has become so central to Western environmental thought, politics, and action. Here, Nature is a single, timeless, and pure domain untouched by Society, or at least the actions of modern humans. This Nature can be known by objective Science and defended and restored by rational environmental management.[6]

In the reading I offer here, the Anthropocene describes a very different world. This world is hybrid—neither social nor natural. It is nonlinear rather than in balance. Futures will not be like the past and will be shaped by human actions. Multiple natures are possible. Science will be complicit in this modification and is political. There are multiple forms of natural knowledge—not all of which are scientific or even human—informing a myriad of discordant ways of living with the world. The result is a proliferation of knowledge controversies. This knowledge politics is unequal and relates to distinct forms of political economy. In short, there is no single Nature or mode of Natural knowledge to which environmentalists can make recourse. The Anthropocene is multinatural.[7]

This account differs markedly from popular approaches to environmentalism emerging after the diagnosis of the Anthropocene. As Paul Wapner explains, these have tended to cleave in two seemingly divergent directions.[8] The first, the "dream of mastery," presents the Anthropocene as an economic and scientific opportunity necessitating more modernization—more knowledge, more technology, and better (i.e., more rational) forms of social and environmental organization. Here, impending disaster legitimates accelerating projects for global science, global markets in ecosystem services, and authoritarian interventions for geoengineering:[9] a final, optimistic modern leap to reconcile humans and the environment under the aegis of sustainable development. This is business as usual for ascendant free-market environmentalism.[10] In Wapner's second direction, the "dream of naturalism," the geology confirms the unnatural character of modern, urban, industrial society. The Anthropocene legitimizes various modes of retreat: renaturalization based on a return to some premodern or even prehistorical state revealed through a valorization of traditional/indigenous knowledge.[11] This is business as usual for modes of deep-green (and generally North American) environmentalism.

In spite of their differences, these environmentalisms have common flaws. They preserve the Nature–Society binary, valuing either "worlds without us" or domesticated environments subsumed to the logics of market exchange. In so doing, they share a totalizing and anthropocentric belief in the power of science and technology to either destroy or manage the earth. This relies on a linear understanding of time, configured around an axis of human progress and decline. The power afforded the Anthropos in these accounts is misplaced and hubristic. It first neglects our persistent vulnerabilities to the earth's unruly geopower manifest in earthquakes, tsunamis, and other geological hazards.[12] Second, it downplays the biopower and resilience of life itself, which continues to elude Promethean aspirations for planetary management and will no doubt survive even the most extreme scenario for a warming world.

These approaches share political flaws. As a growing body of critical work makes clear, scientific invocations of a planet-shaping Anthropos summon forth a responsible species—or at least an aggregation of its male representatives. A common "us" legitimates a biopolitics that masks differential human responsibilities for and exposures to planetary change.[13] It justifies authoritarian governance by a cadre of (largely white, male, and Western) scientists and politicians. It effaces a vast range of alternative ways of knowing and valuing the world.[14] The dream of mastery denies nonhuman claims on the planet,[15] whereas the return to Nature denigrates life forms emergent from and dependent on human care; there is no place for domestic and feral species in the wilderness.[16] This politics is facilitated by a common temporality of impending apocalypse that accelerates action and forecloses on due political process.[17]

These are damning criticisms, fatal perhaps. But the Anthropocene is still a young and immature concept. It has terminological deficiencies and political problems. It has had an awkward genesis, but I don't think it is irrecoverably flawed. In this book I want to harness the potential of its epochal diagnosis to deliver a necessary shock to environmental thought—a shock that has been foretold by a range of critical work in the social sciences that I introduce below. I write about conservation after the Anthropocene in a dual sense: first, at the present juncture after the event of its shocking diagnosis; second, to sketch a future mode of environmentalism for life after the deficient planetary relations the Anthropocene describes.

Here, I am especially drawn to an account of the Anthropocene produced by its conceptual architects, which identifies three periods in the short history and imagined future of the epoch.[18] These first two are familiar and describe the Industrial Revolution (1800–1945) and the subsequent "Great Acceleration" in the processes initiated by industrialization. They argue that a third phase, entitled "Stewards of the Anthropocene," is beginning, as "humanity is, in one way or another, becoming a self-conscious, active agent in the operation of its own life support system."[19] In a rather arbitrary periodization, that is nonetheless convenient for the publication of this book, they suggest that this phase will start in 2015. They identify three future scenarios, the first being business as usual followed by two contrasting modes of stewardship involving mitigation and geoengineering.[20]

This account suffers from many of the problems identified above. Critical social scientists will find it rather too grandiose, technical, and apolitical. But it is heartening for its optimism and ambition and is useful in identifying the present as a key tipping point in planetary governance. Here, I propose an alternative scenario: that the diagnosis of the Anthropocene and the popularization of the "end of Nature" has the potential to value and catalyze modes of "stewardship" based on diverse, reflexive awareness of the always-entangled nature of humans with their environments, the indeterminacy of ecology, and thus, the contested nature of any aspirations toward environmental management—from the local to the planetary scale. Perhaps we could push the zeitgeist for geological epochs a bit further and propose a new epoch after the Anthropocene: the Cosmoscene.[21]

The Cosmoscene would begin when modern humans became aware of the impossibility of extricating themselves from the earth and started to take responsibility for the world in which they lived— turning to face the future, rather than running from the past, and acknowledging, building, and absenting from relations with all the risky, sustaining, and endearing dimensions of the planet. The Anthropocene would become a staging point, the threshold at which the planet tipped out of the Holocene before embarking upon a post-Natural epoch of multispecies flourishing with its own, perhaps less dramatic, stratigraphy.

CONSERVATION

This is a grand, bold promise no doubt beyond the scope of a single book. I explore its potential through a partial and more modest engagement with nature conservation and the governance of the biological dimensions of life on earth. This book is not a synoptic survey of contemporary environmentalism, nor will it have much to say about the "geo" and the wider range of "planetary boundaries" threatened by the Anthropocene.[22] Nonetheless, conservation offers an exemplary domain of environmentalism for my analysis. It is a historic, well-established, and globalizing enterprise well aware of human impacts. It is steeped in Nature thinking and involves science, politics, and practical encounters with life that are characterized by Wapner's dreams of both mastery and naturalism.

Traditionally, and still most commonly, conservation is reactive. It seeks to preserve a fixed Nature from modern, urban, and industrial Society by enclosing it in National Parks. These take the form of prehistorical "wilderness" in North America and much of Africa and South Asia or premodern countryside in Europe. This involves a combination of natural science and romantic iconography. It conjoins aristocratic patronage and state and civil society bureaucracy. Increasingly, though undoubtedly ambivalently, conservation is embracing the market. The past twenty years have seen the proliferation of financial, administrative, and biological technologies for commodifying Nature—from ecosystem services to ecotourism to gene banks. Under the guise of naturalism or mastery, both of these approaches seek control over human and nonhuman life.

My aim in this book is to develop and illustrate a multinatural approach to conservation after the Anthropocene. Its principal contributions are fourfold. I first offer an alternative ontology that conservationists might use in place of Nature. This acknowledges the hybrid and lively character of a world animated by a vast range of human and nonhuman difference adhering to multiple and discordant spatio-temporal rhythms. Second, I present conservation as a set of embodied and skillful processes of "learning to be affected" by the environment.[23] This offers a realist epistemology that attends to the multiple, uncertain, and experimental processes through which natures are known. I examine both in situ encounters in the field and those mediated by the "fingery eyes" of moving imagery.[24]

Third, this leads to an environmental politics that acknowledges multiple forms of expertise and value. Not all of these are human; little is rational or instrumental; and there is frequent discord. I begin to explore the politics of conservation that cannot make recourse to Nature. Fourth, I explore conservation as modes of biopolitics shaping future worlds through the operations of assemblages of scientific knowledge, administration, and practice. These modes have different aims and take place in contrasting political–economic formations. I critically examine a range of contemporary forms of conservation to find a way between the twin poles of mastery and naturalism. I conclude with some positive suggestions for conservation in the Cosmoscene. This is premised on the flourishing of difference, involving the conduct of multiple, often antagonistic, and unpredictable actors and forms of expertise. This book is part critique, part manifesto. It is upbeat and offers constructive criticism to open a conversation with conservation.

The theoretical arguments in the book emerge from and are illustrated by over a decade of research on nature conservation. I draw on the general conservation literature, including scientific papers, policy documents, and grey literature and popular media. I supplement this with data generated through three substantive, interwoven, and ongoing pieces of original fieldwork. Together, these cover a range of important knowledge practices, types of management, and forms of political economy in conservation. My research and, thus, my argument are largely focused on developments in the United Kingdom, continental Europe, and South Asia, but the trends discussed are often global and applicable in other regions.

The first piece of research comprised an investigation of the invention of biodiversity as a new way of organizing international conservation and its arrival in the United Kingdom in the 1990s. It features an overview of the sector and case studies of the conservation of the corncrake and low-intensity agriculture in the Scottish Hebrides and of urban conservation. The second set of materials stems from an examination of international conservation volunteering from the United Kingdom, focusing in particular on Asian elephant conservation in

Sri Lanka and other parts of South Asia. The third project is an investigation of historical and current enthusiasms for "rewilding" and "dedomestication" in European wildlife conservation, with a specific focus on the Oostvaardersplassen—a polder in Netherlands. I provide a more extensive summary of the structure of my argument at the end of this chapter. Before doing so, I introduce some of the key concepts that inform my analysis.

WILDLIFE

To ground a multinatural approach to conservation, I revive and rework the term *wildlife*—a rather antiquated word associated with prebiodiversity natural history. I develop an understanding of wildlife that was first presented by Sarah Whatmore in her influential *Hybrid Geographies*.[25] In wildlife I find an alternative ontology to Nature to inform future environmentalism. An ontology is a theory of what the world is; it establishes key categories, relations, and processes. Wildlife might not seem like an obvious place to start. There is a common assumption that the end of Nature equates to an end to wildness, a domestication of the planet.[26] This is the case only if we accept the mapping of wildlife to wilderness, to places defined by human absence.[27] Instead, wildlife lives among us. It includes the intimate microbial constituents that make up our gut flora and the feral plants and animals that inhabit urban ecologies.[28] Risky, endearing, charismatic, and unknown, wildlife persists in our post-Natural world. Unlike Nature, wildlife also suggests processes. It describes ecologies of becomings, not fixed beings with movements of differing intensity, duration, and rhythm. Wildlife is discordant, with multiple stable states. It is not in any permanent balance. It is shaped by but divergent from the past, multinatural in its potential to become otherwise.

I develop this ontology of wildlife in more detail in the following chapter, which reviews parallel and interdisciplinary developments in the social and natural sciences. I build first from the writings of Bruno Latour, Donna Haraway, and Gilles Deleuze, who in different ways challenge the modern Nature–Society binary and the political settlement to which it has given rise. I engage with a wider literature in which their thought has been developed to offer multinatural and more-than-human grounds for environmentalism. Here, I build in particular on a body of work within my own discipline of geography by scholars such as Sarah Whatmore, Steve Hinchliffe, and Bruce Braun. Their analysis of the problems of Nature precedes the popular diagnosis of the Anthropocene, making comparable observations about the fundamental hybridity and nonlinearity of the planet.

I bring this work into conversation with writings from the conservation sciences. The diagnosis of the Anthropocene has coincided with and energized a period of soul searching, dispute, and realignment within the conservation movement.[29] There is a popular recognition that conservation is failing. In spite of its dramatic growth as a form of governance in the twenty years since the Earth Summit in Rio, biological diversity continues to decline. A growing awareness of the present and future trajectories of agriculture, climate change, and invasive species has led many conservationists to acknowledge the impossibility of saving a pure and timeless Nature. Instead, they focus on the "novel ecosystems" of the Anthropocene.[30] I draw in particular on the excellent account of this new paradigm offered by Emma Marris in *Rambunctious Garden: Saving Nature in a Post-wild World*—though I reverse Marris's terms for what has been lost and what should be saved.[31]

For the later decades of the twentieth century, critical social scientists largely avoided positive articulations of ontology due to concerns with their disciplines' unsavory histories of biological and environmental determinism. These claimed natural causes (e.g., race or climate) for social phenomena (e.g., development), placing them beyond politics. It was sufficient to debunk such claims as social constructions. Ontology was left to the scientists. The result was the realist-versus-idealist (or relativist) impasse that plagued debates between the social and natural sciences in the 1990s. This was true with work on conservation, which reveled in deconstructing claims for authentic Nature and the modes of management they naturalized.[32] But once hybridity had been revealed, this work had little to say about the character, dynamics, or desirability of different material worlds.[33] An ontology of wildlife helps move beyond this impasse, offering a positive, realist, but nondeterministic ontology to inform interdisciplinary science and debate.

WILD EXPERIMENTS

A hybrid and discordant ontology of wildlife has important epistemological and political implications for conservation. Cast off from
the certainties of Nature, how are past and present ecologies known?
How might their futures be predicted? What should be conserved if
multiple futures are possible? Who should decide and through what
processes? To engage these questions I offer an epistemology of conservation as comprising a series of wild experiments—speculative
practices unsure of future outcomes. I first draw on work developing a "more-than-representational" account of knowledge practices.[34]
This approach undermines the Cartesian separation between a rational human mind and an instinctive animal body. It challenges the
prevalent figure of the Human scientist as a "brain-in-a-vat"[35] sensing
the world through disembodied vision and draws attention to the importance of affect—the precognitive sensory mechanisms, perceptual
energies, and feelings that link bodies in encounters.[36]

I develop this work to present conservation as tentative and skillful
processes of "learning to be affected" by a target organism or ecology,
disciplining one's body to tune in to its forms and dynamics.[37] Concentrating in particular on field science, I attend to the embodied,
multispecies encounters through which the flux of wildlife gets sensed,
known, and represented in conservation. I extend this analysis of affect to present conservation as a passionate practice, energized by the
enthusiasms of scientists, volunteers, and other publics in their quest
for valued encounters with other species. Conservation is not rational,
solely motivated by the instrumental desire to secure the delivery of
ecosystem services.

Instead, I identify a range of "affective logics" that frame interspecies
encounters in conservation.[38] An affective logic describes a habituated
mode of engaging with, knowing about, and feeling toward wildlife.
These are cultural phenomena emergent from bodily encounters. I
configure my analysis of affect in conservation around a discussion of
nonhuman charisma. Conservationists frequently talk about the key
roles played by charismatic species, but this charisma remains under-
theorized. I develop a tripartite understanding of charisma that moves
out from the anatomical and ecological properties of the nonhuman in

question to explore the aesthetic dimensions of interspecies encounters with both proximal and distant conservation publics.

These encounters between conservationists, publics, and other species do not occur in a political vacuum. To contextualize these knowledge practices, I figure conservation as proceeding within an assemblage.[39] The concept of an assemblage describes the "stuff of politics": the material ecology of bodies, technologies, texts, and other materials through which knowledge is produced and ordering takes place.[40] The assemblage of conservation is heterogeneous. In addition to lively human and animal bodies, it comprises nature reserves, fences, and guns; scientific instruments, maps, papers, and databases; legal designations, action plans, and market mechanisms; and films, websites, and online transfers, to give a few examples.

Assemblages allow certain actors to speak for, commodify, govern, and thus shape the world, often in conflict with other representations. Assemblages have inertia. They are haunted by pasts, groove present practice, and serve to anticipate different futures. Assemblages have geographies that perform connections and link bodies and places in multiple spatial, or topological, formations. Assemblages allow elites to act at a distance. Assemblages are always partial and dynamic. They are under way and on the move. The concept of assemblage seeks to convey process, and when conceived as a process, any assemblage is thus potentially unstable. No assemblage is hegemonic.

This approach offers a multinatural epistemology that recognizes multiple ways of being affected by the world, encoded in a range of affective logics incarnated in material assemblages. I develop this approach to explore the knowledge practices of conservation as experimental. Rather than seeking to test explicit theories and hypotheses framed by transcendent archetypes of Nature, these experiments involve an open-ended set of practices likely to generate surprising results. Here, an experiment is a trial or a venture into the unknown.[41] In the field they often involve deliberations with numerous publics and forms of expertise in situations where multiple futures are possible and there is no clear division between lab and field.

Michel Callon and his fellow researchers have explored the various techniques through which publics can be involved in such multinatural experiments.[42] They differentiate "research in the wild" from "secluded research." The latter, they argue, is most commonly associated with the lab (though it can take place in the field) and has tended to cut itself off from the publics it subsequently affects. Such secluded research still has an important role, but they argue it should be linked to its publics through engaging in research in the wild among emergent collectives of expertise.

Research in the wild implies neither a disavowal of nor a eulogy to science, bureaucracy, or reason. It helps to bring them into politics. As Jane Bennett has compellingly shown, *wildness* can mean more than thought from outside civilization—the romantic residual in reaction to the alienation of modern life.[43] In keeping with recent reevaluations of the term, I propose an epistemological and political place for wildness at the heart of contemporary life.[44] Here, wildlife is vernacular, everyday, and democratic.[45] It provokes curiosity, disconcertion, and care. It demands political processes for deliberating discord among multiple affected publics.

We can think of the wild as the commons, the everyday affective site of human–nonhuman entanglement. Politics in the wild involves democratizing science, relinquishing the authority that comes with speaking for a singular Nature. Multispecies, often urban, wilds are where political life takes place now that the laboratories of modern science have taken over the world and we have all become caught up in the global experiment that is the Anthropocene. I hope to show how political–ecological experiments in such wilds offer new ways of conceiving and practicing environmental politics and of living with human and nonhuman difference.

In my analysis of the political ecology of conservation, I am especially concerned with the different values placed upon encounters. I draw on and develop Donna Haraway's brief discussion of encounter value and the emergence of forms of "lively capital."[46] The concept of nonhuman charisma helps to develop a taxonomy of encounter value. I engage with the extensive literature on the political ecology of conservation to explore the different ways in which encounters get valued under different modes of political economy. I attend in particular to the commodification of encounters in spectacular modes of neoliberal conservation, identifying the power of commodified flagship species in funding and framing conservation action.[47] I examine the implications of commodified encounters for the animals, ecologies, and marginal publics subject to this form of conservation practice. Thinking of conservation as wild experiments means giving up on Nature. This is risky. A fixed Nature, known by Science, is the fulcrum for the territorial, legal, and political gains made by the conservation movement during the twentieth century. A hybrid and immanent ontology could be more conducive to the demands of neoliberal capitalism than a fixed Nature.[48] Multiple fluid natures are perhaps more fungible and amenable to the logics of market exchange. And the recognition of multiple forms of environmental expertise risks undermining the authority of Natural Science—generating "skepticism" and facilitating discord to preserve the status quo.[49] Though Emma Marris has little to say about the politics of her "rambunctious garden," her book and the wider interventions proposed by the "modernist green"[50] movement in conservation has prompted debate, disquiet, and nascent changes within the sector.[51] These are important issues that I take up in this book and discuss at length in the conclusion.

BIOPOLITICS AND COSMOPOLITICS

To trace the operations and significance of these experimental encounters, I present conservation as a type of biopolitics, where biopolitics describes a modern form of governance that seeks to secure the future of a valued life (both human and nonhuman) at the scale of the population. Biopolitics involves the systematic, but never totalizing, application of scientific knowledge, technology, and administration. I am particularly interested in how different modes of conservation come to shape different worlds, cutting up the flux of wildlife and performing particular ideas of what life should be saved. I term this process "ontological choreography" after Donna Haraway.[52] I explore how different modes of conservation cut up wildlife according to different knowledges and in the interests of different human and nonhuman actors. Tensions between modes of conservation result in what Anne-Marie Mol has termed "ontological politics," whose outcomes become vital in shaping the planet in the Anthropocene.[53] In short, my argument is that conservation after the Anthropocene is performative, actively shaping subjects and ecologies in relation to the knowledge by which it is informed.

In this book I develop a rather eclectic approach to biopolitics, tailored to an analysis of the science, politics, and practice of conservation. Its basic principles are drawn from the work of Michel Foucault, who identifies two "political strategies"[54] original to modern forms of government.[55] The first, which he terms "governmentality," describes the rise of powerful knowledge practices that construct standardized models of normal, rational, healthy citizens and inform technologies that discipline individual adherence to these subjectivities. Arun Agrawal has reworked this concept to describe the "environmentality" of conservationists' efforts to create environmental human citizens.[56] I engage with and develop this work to explore how the affective logics of conservation are governed through the mediation and commodification of conservation encounters, under different forms of political economy.

My main interest is in the second political strategy Foucault identifies. This is the emergence of modern forms of "biopower" where the concern shifts from the behavior of individuals to the management of life at the scale of the (often unruly and unpredictable) population. Foucault highlights how modern "biopolitics" involves productive and destructive processes through which life is made to live or left to die. The concept of biopolitics is now commonplace in the social sciences and informs critical analysis of the deployment of natural science to manage populations to secure human and environmental health.[57] Foucault is resolutely human in the foci of his analyses of biopower and notoriously ambivalent about animals and the environment as political problems.[58]

Post-Foucauldian scholars have developed the concept of biopower to identify and analyze the multitude of modes of nonhuman biopolitics that characterize late-modern governance.[59] Perhaps the most well known are Giorgio Agamben's writings on the "anthropological machine": the categorical procedure through which lines are drawn between human, political life (bíos) and bare, animal life (zoē).[60] Agamben is most concerned with the deadly consequences for humans of being rendered animal—what Foucault termed "thanato-politics"[61] and has little to say about the effects on nonhumans of being rendered zoē.[62] This work is important, but as various critics have argued, it is rather too totalizing, anthropocentric and, deathly. It presents biopolitics as the control over life and neglects both the generative dimensions of securing life and the ability of life to do otherwise.[63]

In this book I draw on and develop a range of livelier and more affirmative approaches to biopolitics that are willing to afford some power to the "bio." One key source is Haraway, who presents biopolitics as processes of "living with"—modes of companionship figured as unequal and power-laden but nonetheless contingent and more-than-human dances of relations through which material bodies learn to be affected by one another.[64] Her approach is more consistent with the ontology of wildlife that informs this book. It culminates in an appeal for a "cosmopolitics"—a concept she takes from Isabelle Stengers—premised on the flourishing of multispecies difference.[65] Although she is interested in the biopolitics of breeds and species, Haraway takes the individual organism (largely dogs) as the unit of her analysis.

Haraway's cosmopolitics of living with resonates with work in geography by Steve Hinchliffe, Sarah Whatmore, and their co-researchers on the biopolitics of biosecurity (the governance of mobile plant and animal disease) and urban conservation.[66] This work has focused more on processes, landforms, and species less familiar to humanist models of biopolitics. For Hinchliffe a cosmopolitics for living with aggregate nonhuman populations involves anticipating, nurturing, and managing events that emerge from the circulation of human and nonhuman actors in diverse spatial formations (or topologies).

This cosmopolitics is not about rendering the present eternal but involves a careful, processural political ecology that is open to the immanent "likely presences" of nonhuman life.[67] Epistemologically, it is aligned with (and informs) the concept of wild experiments outlined earlier. Their work is founded on a political commitment to "putting accepted knowledges at risk" by working with emergent collectives of experts, not all of which are human. It offers a science-politics that does not make resource to Nature.[68]

These approaches to biopolitics as processes of living with nonhumans figure conservation as tentative processes of working with the biopower of the ecologies and organisms that comprise the nonhuman world. They require a humble, less anthropocentric model of the ontological choreography of conservation. Not only is conservation marginal in relation to other human claims on the earth, but also it is rarely in control of its target ecologies. Beyond the limited set of organisms nurtured by agriculture, the species that are faring best in contemporary hybrid ecologies are those most able to occupy its modified spaces and spatialities. Invasive "global swarmers" trouble conservationists as biosecurity threats, pest species that threaten biodiversity and circumvent human efforts toward their control.[69] We should be wary of the popular anthropocentric metaphor of biodiversity conservation as an ark for the Anthropocene. The biopolitics of biodiversity will shape but not determine future ecologies. There are powerful inhuman natures at work here on a dynamic and warming planet that will shape future ecologies.

In his writings on governmentality and biopolitics Foucault is concerned especially with the rise of neoliberalism. The ascendance of this mode of political economy has troubled much subsequent writing on biopolitics as well as critical work on nature conservation. Neoliberalisms are less central in the accounts that follow, as they are less significant to the forms of conservation about which I write. Conservation in Europe during the period I describe was dominated by a range of nongovernmental organizations working in conjunction with sympathetic statutory authorities at the national and European scale. These groups were largely opposed to the logics of private property, markets, and commodification. Conservation management was funded through volunteer donations, direct public payments, and most significant, taxpayer-funded agro-environmental subsidies delivered through the EU Common Agricultural Policy. This is changing and is certainly less the case with the market-oriented modes of conservation that I discuss in chapter 7.

STRUCTURE OF THE ARGUMENT

The first three chapters outline and illustrate the conceptual foundations of the approach to conservation I have summarized. In chapter 1, I present the ontology of wildlife that forms the foundations for this book. I illustrate this with reference to Asian elephants and the political ecology of Sri Lanka. In chapter 2, I explore conservation as a process of learning to be affected and present the concepts of non-human charisma and affective logics in conservation. I explore these through a set of reflections on bird surveillance. Chapter 3 offers my first take on the biopolitics of conservation. Here, I trace the arrival of biodiversity as a new way of understanding and governing wildlife in the United Kingdom. I focus in particular on the scope of what gets understood and conserved, identifying a distinct taxonomy that maps onto the forms of nonhuman charisma identified in chapter 2. I trace the performance of this oligopticon and reflect on the role of the material assemblage of conservation in shaping this mode of biopolitics.

The next two chapters detail and compare prevalent modes of contemporary conservation. In chapter 4, I offer a detailed case study of corncrake conservation. The corncrake is a rare and threatened migratory bird that inhabits the marginal landscapes of the Scottish Hebrides. It is dependent on the preservation of crofting, the local low-intensity agricultural system. I explore the biopolitics of corncrake conservation, tracing how the corncrake was aggregated as a dynamic population modeled to calculate optimum modes of corncrake-friendly land management. I reflect on how corncrakes and crofters were governed through these interventions. I take the corncrake as exemplary of a mode of conservation biopolitics that I term "conservation as composition." This is targeted at species and is rooted in equilibrium ecology. It seeks to render the present eternal—subsidizing, deliberating with, and regulating human land uses to prevent both intensification and abandonment.

In chapter 5, I compare this mode of biopolitics to a very different form of conservation associated with the recent enthusiasms for rewilding. I provide a critical analysis of a flagship example of this approach in the management of the Oostvaardersplassen—a polder in the Netherlands. Rewilding shifts the historical benchmark of conservation to premodern landscapes and focuses on the restoration of ecological processes. It advocates land sparing rather than land sharing. In some cases it involves experimental, open-ended forms of science and management less sure about what an ecology might become. It is controversial, not least because it challenges the science and policy associated with the compositional model of conservation. It is risky in its appeal for nature development. Through a critical and affirmative analysis of this example, I identify the promise and risks of this alternative.

The final three chapters focus on a significant domain within the biopolitics of conservation. Chapter 6 examines moving imagery and the affective logics that characterize wildlife film. Conservation depends heavily on media for fund-raising, advocacy, and education. Many of us live in media ecologies in which we are more likely to encounter rare and charismatic wildlife in screen than in the flesh. Returning to elephants, I critically examine four prevalent logics according to which animals are evoked and reflect on their implications for modes of environmentality shaped within media ecologies. I identify the potential of curiosity as an affective logic for attuning to the difference of wildlife.

In chapter 7, I look at markets and explore one mechanism through which wildlife is brought to the market in contemporary conservation. I focus on the commodification of valued encounters with charismatic species. I develop the concept of nonhuman charisma introduced in chapter 2 and trace its increasing significance to emerging and powerful forms of spectacular neoliberal conservation. Focusing largely on Asian elephant conservation, I reflect on the biopolitics of configuring conservation around commodified encounters. I examine the implications for individual captive animals, wider ecologies, and the marginal farmers forced to live in proximity with free-ranging members of this charismatic flagship species.

Chapter 8 turns to questions of space. It explores the geographies of wildlife through a critical analysis of the topologies associated with different approaches to wildlife conservation. Topology is a branch of mathematics that invents new ways of conceiving spatial relations beyond the familiar cartography of the topographic map. I explore how thinking topologically helps identify the territorial trap into which modern conservation fell when it configured biogeography around purified nature reserves. I examine how this regional topology has been challenged first by urban conservationists and then by the connectivity turn that is currently taking place in conservation. I explore the utility of networks, fluids, and fire as alternative topological metaphors for examining the biogeographies of wildlife conservation after the Anthropocene.

The conclusion returns to the broad aims I outline at the start of the introduction. It summarizes the cosmopolitics of wildlife conservation that I develop and gives an overview of my aspirations for conservation after the Anthropocene, distilling the contributions of this book to engaging in some "anticipatory semantics" with the concept of the Anthropocene. I finish by identifying some tensions within and challenges to this model. I discuss the ontological politics of conservation, the interface between wildlife and biosecurity, and the relationships between wildlife conservation and neoliberal capitalism.

· 1 ·

WILDLIFE

Companion Elephants and New Grounds
for Multinatural Conservation

There are between three and four thousand Asian elephants living on
the densely populated island of Sri Lanka in the Indian Ocean (Figure 1).[1] They are intelligent, emotional animals who live long lives in
complex social groups. Given the choice, they would be wide-ranging,
tramping extensive territories along established lines of movement.
At present they inhabit a fragmented and dynamic biogeography
comprising protected areas, cultivated land, orphanages and transit
homes, and various modes of private and religious captivity. There is
a long history in South Asia of taking elephants from the wild and
training them for work and ceremony.[2] Recent scientific research confirms historical accounts of the international trade in elephants and
their frequent movement between free-ranging and captive spaces.[3]

Elephants rarely breed in captivity and are not considered to have
been domesticated,[4] but contemporary animals regularly interact with
people and show great behavioral plasticity in adapting to different environments.[5] Free-ranging animals have flourished in active and abandoned agricultural landscapes and would continue to do so were it not
for the conflicts this generates with farmers.[6] Instead, elephants and
people in Sri Lanka are enmeshed in a lethal, multispecies ecology in
which hundreds of humans and elephants are terrified, maimed, and
killed every year. For conservationists these elephants are an important population of the globally threatened species *Elephas maximus,*
and the Asian elephant has been promoted as a flagship species for
regional conservation.[7]

FIGURE 1. A herd of Asian elephants at the Pinnawala orphanage in Sri Lanka.
Photograph by Bernard Gagnon, Wikimedia Commons.

Sri Lanka's elephants and their habitats are not easily understood
as Nature. They are too social and sagacious to be objects, too strange
to be human, too captive and inhabited to be wild, but too wild to
be domesticated. There are multiple natures at play in these ecologies
and valued ways of being that are more-than-human. There are long,
fraught histories of interspecies exchange that precede the originary
moment of the Anthropocene and trouble its epochal status. In this
chapter I start with these elephants in order to outline an alternative ontology of wildlife that environmentalists might use in place
of Nature for conservation. I take inspiration from their wild un-
Naturalness to illustrate ways of moving on from Nature. I hope to
offer more positive, multinatural grounds for conservation thought
and practice. This chapter introduces some of the key concepts that
guide this book. It should be read in conjunction with the one that fol-

lows, which performs a similar service to the category of the Human Scientist.

ONTOLOGY

An ontology is a theory of what the world is; it establishes key categories, relations, and processes. I draw my inspirations for an ontology of wildlife from a range of sources. My principal touchstones are in the social sciences and comprise a set of multinatural and more-than-human approaches that stem primarily from diverse engagements with the social theory and biophilosophy of Bruno Latour, Gilles Deleuze, and Donna Haraway. Jane Bennett has termed these approaches modes of "vital materialism."[8] Such thinking has been developed within my own discipline of geography by figures such as Sarah Whatmore, Steve Hinchliffe, and Bruce Braun, and my approach is indebted to their pioneering work.

My second source of inspiration is the natural sciences, where (albeit still marginal) developments in disciplines like ecology, ethology, and conservation biology offer un-Natural approaches that resonate with these vital materialisms. There is important common ground emerging between these broad fields. In the sections that follow, I summarize these affinities in relation to four main themes: *hybridity, nonhuman agency, immanence,* and *topology.* What follows is an explicitly normative, synthetic, and optimistic account. It has its problems, some of which I anticipate in conclusion.

Hybridity

Research on the environmental history of Sri Lanka suggests that much of the island is covered in secondary forest that grew back after the collapse of various premodern civilizations.[9] This ecology of abandonment was more conducive for Sri Lankan elephants than the dense forest it replaced, and their populations flourished. This work challenges orthodox understandings that figure the nation's forests as untouched wilderness and suggests that these habitats and their denizens are hybrid. Here, *hybrid* describes a mixture comprising parts of two (or sometimes more) forms—in this case the Natural and the Social.

As I explain in the introduction, hybrid ontologies are very much in vogue in the social sciences in the wake of a series of conceptual critiques of the prevalent modern Nature–Society dualism. Two of the most famous are those offered by Bruno Latour and Donna Haraway in the early 1990s.[10] These authors argue that the dualism is the result of the translation and purification of the world by Natural Scientists, not the revelation of its transcendent essences.

Hybrid thinking has had a great influence on a body of research concerned with the wider place of wilderness and other dualistic ontologies in conservation.[11] Wilderness offers a powerful and influential imagination for various modes of (largely colonial and postcolonial) conservation. It imagines a pure and ahistorical place for Nature— natural by virtue of being untouched by human hands. It celebrates wild animals most distant from domestication. In a famous essay on "the problem with wilderness," the environmental historian William Cronon discloses the ontological impossibility of wilderness and its political and ecological problems as a category for conservation policy.[12]

The imagined purity of wilderness is less significant in European conservation, where the valued baseline tends more toward the premodern than the prehistoric. Here, Nature is located in a past countryside produced through various naturalized forms of low-intensity agriculture. In both cases the nineteenth-century moment of the fall is consistent, as is their diagnosis of the problem: the advent of modern, urban industrial capitalism across western Europe and North America. Modern people have no place in Nature.

The research in the forests in Sri Lanka can be situated within a growing interest in the natural sciences in hybrid forms and spaces. Palaeoecologists have traced the extensive signatures of past human activities in conservation territories popularly figured as pristine.[13] In a similar fashion, a growing body of research in contemporary conservation biology examines the "novel ecosystems"[14] or "anthromes"[15] emerging in the Anthropocene. In their work on "countryside biogeography," Gretchen Daily and her colleagues note a growing recognition that, "from both purely academic and practical perspectives, ecologists should be able to say more than 'weedy' about the biota that may survive human impacts."[16] Michael Rosenzweig has proposed a new mode of "reconciliation ecology" that "discovers how to modify and diversify anthropogenic habitats so that they harbor a wide variety of wild species."[17] This thinking is influential in the ambitious field of ecological restoration and design.[18] The focus here is on the area conservationists tend to refer to as the matrix—inhabited land around nature reserves that is commonly derogated as lost. Emma Marris offers a compelling review of these developments in *Rambunctious Garden*.[19] An interest in hybridity also characterizes a rich body of recent writing in animal studies concerned with the nature of specific organisms and interspecies relations that confound any simple human–animal or wild–domestic dualisms.[20] The aim here is to open up the category *animal*, to recognize the multiple forms of difference it subsumes, and to take seriously the ways in which humans and animals are shaped by their interactions.[21] This is a growing and vibrant field that has been championed by Donna Haraway.[22] Building from her own ways of living with dogs, Haraway traces the entangled relations through which canines and humans have coevolved. She offers a hybrid ontology of "companion species."[23]

We can understand Sri Lanka's elephants as archetypal companion species and can trace their diverse entanglements within multispecies histories and geographies. Elephants have coevolved with people over millennia. Their genetics, anatomies, behaviors, feelings, social groupings, and wider ecologies all bear a human signature. At the same time, the language, culture, religions, agriculture, and economies of their human coinhabitants carry a pachyderm trace. These relations are unequal and frequently fraught and cut across species divides.[24]

A companion species ontology thus blurs the Social end of the Nature–Society dualism, challenging the ontological security of the Human. Haraway extends Latour's famous claim that "we have never been modern" to argue that "we have never been human."[25] She traces the lively materialities of interspecies interaction—including genetic, microbial, haptic, digestive, and ecological connections—to demonstrate the ontological impossibility of extracting a bounded and uniquely human body from the messy relations of the world. This work draws attention to parts of the biological kingdom that are poorly known and have rarely concerned conservation biologists.[26] For example, she flags recent research that reveals the nonhuman composition of human (and other animal) body and our microbiotic interdependence with other species.[27] We are what we eat, drink, and breathe, and these actions shape wider ecologies. We are bitten, infected, and parasitized by diverse microbial organisms. The historical entanglements of human and elephant bodies can be traced at this microbial scale. Epidemiologists report a multitude of zoonotic viruses and bacteria that cross between humans and elephants.[28] One of the most promiscuous is *Mycobacterium tuberculosis* or TB. Humans act as reservoir hosts for this pathogenic bacterium, which can be passed to elephants by infected water droplets. Research on elephant paleoepidemiology has suggested that TB was pandemic among mastodons (distant ancestors of *Elephas maximus*) and may well have triggered a "hyperdisease" event.[29] Human hunting and interspecies infections may have led to the late-Pleistocene extinction of several proboscidean species (including the mammoth).[30] Contemporary descendants inherit this immunology and regularly host TB in captivity,[31] though veterinarians across South Asia are concerned about the risk of new antibiotic-resistant varieties of TB decimating fragile free-ranging populations.[32]

Thinking through this "corporeal generosity"[33] presents a radically different ontology of the human (and other animals) as supraorganisms, systems composed of multiple organisms. This microbiological ontology is a key element of Lynn Margulis's radical theory of symbiogenesis, which traces the "extreme genetic fluidity" of bacteria and their promiscuous capacities to "merge transiently or permanently with larger organisms."[34] This understanding challenges prevalent ontologies of animal studies that tend to be fixed on beings that are "big-like-us"[35] and flags the significance and radical alterity of the microbiome. Kathryn Yusoff takes this microontology a step further in her recent writings on geology, emphasizing the mineral foundations of all life and the ways in which the bio and the geo are entangled through interdependent webs of biochemical exchange.[36] This work rarely deploys the term *hybridity,* which seems to suggest the preexistence of forms that get mixed. There is little antecedent purity in this model in terms of Nature and Society that can be *reconciled* or *coupled,* to use two popular terms.

Much of the analysis in the environmental social sciences that is informed by these diagnoses of hybridity has tended toward critiques of claims of authenticity. Latour and Haraway's earlier work allows us to trace Nature as a relational achievement, a power-laden process of purification rather than the revelation of a transcendent archetype. This approach flags the paradox—identified by both Cronon and Latour—that late-modern conservation projects that appeal to a prehistorical or premodern Nature are dependent on the very technologies they purport to absent. Comparable critiques in animal studies revel in divulging the nonhuman constitution of the human, flagging the impossibility of specifying a human essence and noting interspecies commonalities. This style of authenticity critique has had some purchase but has ultimately limited utility for orienting wildlife politics. Diagnoses of hybridity swiftly become banal. Yes, the world is hybrid, but how do these hybrids work?

Nonhuman Agency

To answer this question, we need to acknowledge the work done by nonhumans and to attend in more detail to the types of actor and agency that are let loose by a hybrid ontology. This move challenges the privileged place of the rational human subject as the sole locus of agency in humanist accounts of social and environmental change. It draws attention to the role of a range of nonhuman actors, caught up inside and outside relations with humans, that shape the form and dynamics of any ecology. These nonhumans include (and blur the boundaries between) technologies, human bodies, and other biological and geological forms and processes.

The idea of nonhuman agency has become a popular concern within the social sciences as a result of a range of conceptual developments, which I review in this section. These are concerned with different types of nonhumans and their respective agencies. What they share is a disavowal of the (recently somewhat reinvigorated) retreat to determinism that can be detected in some environmentalist accounts of the Anthropocene. These figure Nature as an external force—an apocalyptic avenging angel or essential human essence—that determines the future and thus negates politics.[37]

One of the most influential articulations of nonhuman agency in the social sciences is actor–network theory (ANT), an approach developed in the 1980s and 1990s by sociologists of science like Bruno Latour, John Law, and Michel Callon.[38] ANT offers a "flat ontology"[39] based on the principle of a generalized symmetry between the agencies of human and nonhuman "actants." In place of a binary world of social and natural entities, Latour and his coauthors present a world of actor–networks—entanglements of people, technologies, and other nonhumans with diverse and distributed agencies. The quasi-ecological ontology of the actor–network is now more commonly described by the term *assemblage*, which I explain in the introduction. In practice, ANT is concerned largely with the agency of technologies and other anthropogenic artifacts in permitting action and ordering at a distance. As I illustrate in chapters 3 and 4, this approach provides useful resources for tracing the material connections through which biopolitics takes place.

In spite of their early appeals to ecology,[40] Latour and Callon are skeptical about specifying different types of nonhuman agency within and across their case study inquiries. The human and technological foci of their accounts mean that unruly geological and biological actors do not feature prominently. Nor do they explore the dynamics of lives and worlds where humans are bit players. In short, ANT has struggled to account for the great diversity of "political matter."[41] Fortunately, it has morphed into and showed the way for subsequent strands of more-than-human thinking, many of which have emerged from interdisciplinary conversations with earth, life, and material scientists.

For example, a range of "multispecies ethnographies"[42] in animal (and more recently plant) studies has examined the biopower of nonhumans and the ways in which they sense and shape their worlds.[43] Comfortable with the notion of nonhuman subjectivity, this research has developed modes of "critical anthropomorphism"[44] for thinking like specific organisms and for witnessing and evoking nonhuman ways of being in the world. Here, there is a growing interest in how the material and ecological properties of particular organisms come to shape human–nonhuman relations. This work has tended to focus on prevalent relations with domesticated plants and animals[45] and could be fruitfully connected with cognate science exploring "animal cultures"[46] among free-ranging populations inhabiting the novel eco-systems of the Anthropocene.[47] In the following chapter, I offer a more sustained introduction to and illustration of multispecies ethnography through an analysis of nonhuman charisma and begin to explore some of its implications for wildlife conservation.

Multispecies ethnography helps make sense of the entangled lives of people and elephants. Research on young African elephants ex-posed to poaching and culls suggests that these animals suffer both grief and trauma comparable with posttraumatic stress in human communities.[48] This results in depression and rage, causing elephants to kill each other and the rhinos with which they share National Parks. In Amboseli in Kenya, this animosity extends to humans. Here, elephants can differentiate antagonistic Masai and will target their cattle.[49] Similar problems afflict rogue elephants in Sri Lanka— lone males who have come into conflict with people and then make nocturnal forays to "raid" the crops, garbage, and houses of farmers, frequently killing those they encounter. The sources of this aggression are still unclear but could be attributed to similar trauma.

Haraway and many of her fellow authors in animal studies have tended to take the animal organism as the principal locus of nonhu-man biopower. A different strand of work shifts scales to focus on the molecular composition and dynamics of the bio and the geo. Here, there is a resolutely "inhuman" concern with processes that need not necessarily relate to animal bodies.[50] Recent interest in this approach arises in part from a dissatisfaction with the accounts of nonhuman agency offered by Latour and Haraway. For example, in his writing on geology and "geopower," Nigel Clark accuses Latour of anthropo-centrism. He argues that his account of hybridity ascribes too much agency to humans and neglects the "radical asymmetry" and "indif-ference" of geological processes with which humans have little relation (at least in terms of control) but that have earth-shattering implica-tions for social life. He develops a similar argument in further writings on the unruliness of the bio in the context of invasive species. Clark aligns Latour's approach with the hubristic accounts of planetary management in the Anthropocene that I discuss in the introduction[51] and argues for a more humble appreciation of the uncertain and undomesticated geological and biological contributors to our risk society.[52] This articulation of bio- and geopower is indebted to the philosophy of Deleuze and Guattari. This work offers an inhuman ontology of wildlife figured as a collection of becomings characterized by flows above, through, and below the level of the organism, in ecologies that need not feature a human subject.[53] This philosophy clearly influenced the materialisms of Latour and Haraway, but they have both moderated its far-reaching commitments to immanence.

Immanence

In the context of biophilosophy, immanence describes an ecological assemblage composed of a single substance and characterized by emergent properties, rather than transcendent essences. It suggests a speculative and multinatural ontology, sure of the existence of matter but perpetually uncertain as to what that matter might become. A concern with immanence implies a more explicit concern with process, rather than form, and speaks to the growing interest in time in cognate parts of the ecological and social sciences.

A common aim in this work has been to think beyond linear, cyclical, reversible, and orderly temporalities to offer a range of concepts that examine the creative, nonlinear, irreversible, and open-ended nature of time.[54]

This offers a theory of time without linear trajectories of progress or decline (e.g., from a premodern to a modern epoch). It also offers an approach to nonhuman agency that is nondeterministic, depicting a discordant world composed of a multiplicity of forces and trajectories with the perpetual potential for differentiation. Such a contingent and plural conception of natures promises neither salvation nor apocalypse. Contingent natures are multiple. Not only are they perceived in multiple ways, but they come into being through multiple trajectories. Such contingency precedes politics but by no means determines it.

As several authors have noted, Deleuze's concern for immanence draws on and resonates with the long-standing interest in nonequilibrium, complexity, and uncertainty across the physical and life sciences.[55] In ecology and biogeography, ideas of nature in balance have long been challenged by nonequilibrium ecology. This paradigm is concerned with forms and processes with multiple and often divergent trajectories, wherein single events can have significant and unforeseeable consequences. It describes ecologies with thresholds, tipping points, and multiple equilibria with differing degrees of stability.[56] In conservation biology nonequilibrium thinking is evidenced in recent work on "landscape fluidity" or "the ebb and flow of different organisms within a landscape through time."[57] Emerging in the context of a growing awareness of ecological adaptation to climate change, this work eschews models of linear succession and categorizations of stable climax communities to argue that landscapes are forever "chasing moving targets"[58] and do not stand still. This requires "dynamic reserves"[59] able to adapt to future becomings. Daniel Botkin offers a useful musical metaphor of ecologies comprising "discordant harmonies" without transcendent order, with multiple possible futures.[60] By extension the conservationist as conductor seeks not to reveal the composer's transcendent score but to offer one of many interpretations.

We can see an illustration of this thinking in relation to Sri Lanka's elephants. In 2007 the country's Department of Wildlife Conservation adopted a new approach to land-use planning that departs from the orthodox binary spatial logic of protected area management.[61] Rather than confining elephants through fencing and translocation, it maps dynamic reserves that acknowledge the hybrid environmental history of the Sri Lankan elephant habitat. In these zones they encourage experiments in "temporal resource partitioning," whereby elephants can graze outside National Parks during the fallow season for neighboring shifting cultivators.[62] This reconciliation ecology necessitates a willingness to learn from elephants and the various ways in which they adapt to anthropogenic landscapes.

Topology

Topology is a branch of mathematics which imagines different kinds of space. In particular, it invents spaces by thinking up different rules for defining the circumstances in which shapes will change their form or not. It is possible to devise indefinitely many rules for shape invariance.[63]

Thinking space in terms of topology provides new means of mapping connectivity beyond the Cartesian coordinate system associated with the familiar topographic map. It opens analysis to the multitude of spatial relations performed by different actors and processes entangled within diverse ecological assemblages. Topology has had a great influence on recent work in geography.[64] Geographers have focused in particular on the topology of the network and the ways in which networks depend upon and start to dissolve familiar territories.[65] Networks link people, organisms, and places in ways that confuse cartographies configured around a nested set of regions and nations. As I explore in more detail in chapter 8, topological investigations of urban wildlife and invasive species challenge binary geographies that efface nonhuman life from urban areas or map it exclusively to national territories. This work questions the utility of spatial categories like alien/native, in situ/ex situ, and wild/domestic[66] and acknowledges the more-than-human spatialities performed by people, plants, and animals in "living cities"[67] and global networks.[68] The recent proliferation of topological metaphors helps analyze the biogeographies of an immanent ontology of wildlife in an increasingly mobile and globalized world.

Thinking in terms of networks helps make sense of the un-Natural geographies of Sri Lanka's Asian elephants. Recent work documenting the geographical distribution of genealogical lineages (or phylogeography) of the Asian elephant has identified two distinct groups (or clades) within the wider population. These are thought to result from its contraction to and speciation within separate glacial refugia in Myanmar and southern India–Sri Lanka during the Pleistocene.[69] Because Sri Lanka returned to being an island after the glaciers retreated, scientists were surprised to find that individuals representing both clades were currently in residence. This distribution allowed them to cautiously suggest traces of the long and well-documented history of international elephant trade and subsequent escape. This network topology challenges national territorializations of Sri Lanka's elephants,[70] figuring them as immigrant and feral organisms whose ancestors originate in southern Myanmar. Subsequent research suggests that these animals have different cultures and tend to keep to themselves, at least when choosing to breed.[71]

An interest in network topologies also characterizes the connectivity turn that is currently under way in the theory and (to a lesser extent) the practice of conservation biogeography.[72] This is happening as a result of the shift to the nonequilibrium models of ecology I outline and a growing awareness of the diverse ecological adaptations of climate change. Connectivity is a multifaceted concept, configured by the nature of what is being connected and the spatial and temporal scope of analysis. In its most straightforward framings, it provides an index of spatial linkage and informs demands for "conservation corridors" and "ecological networks" to link together protected areas.[73] Looking beyond protected areas, conservationists advocate forms of landscape permeability in ways that challenge the fixed and purified territories for Nature associated with orthodox protected areas. More hands-on forms of connectivity management promote practices of "assisted migration," translocating and reintroducing organisms that are unable to move or for whom such networked ecologies would be too expensive.[74]

The flip side of connectivity is invasion. Deploying spatial terms similar to those promoting connectivity, environmental historians and biogeographers have begun to trace how the space–time compression associated with globalization has created a "New Pangaea,"[75] a networked biogeography that effaces continental boundaries, links isolated island biogeographies, and reorganizes the conditions in which life has and will evolve. A cosmopolitan flora and fauna of "global swarmers"[76] are proliferating and flourishing in the novel ecosystems of the Anthropocene.

For some conservation biologists, the Anthropocene is also a "homogecene" in which ecological globalization results in the convergence of emerging novel ecosystems and the erasure of established forms of genetic, species, and habitat diversity.[77] Dissenting voices suggest that the forms and relations emerging in novel ecosystems might be more differentiated. Here, ecological globalization might also act as a tool for differentiation producing new hybrids and nurturing threatened natives.[78] As with connectivity, this debate centers on the location and nature of the difference being compared and the scale at which the comparison takes place.[79]

Discourses of connectivity and invasion have been extremely

influential in debates over Asian elephant conservation. Elephants are increasingly confined to fragmented pockets of protected land, their historical pathways of movement cut off by human infrastructure or generating conflict with agricultural land use.[80] Ambitious plans have been drawn up in India to manage the metapopulation inhabiting these fragments through a national network of elephant corridors and the possible consolidation of elephant populations in larger reserves.[81] As I examine in chapter 7, these plans have been controversial. Meanwhile, the recent push for elephant ecotourism to fund these ventures opens new vectors for invasive pathogens, like the TB virus.

To understand connectivity and invasion, we need to take seriously the specific topologies performed by the nonhumans concerned. The nature and degree of connectivity of a landscape relates to the behavioral and/or ecological properties of that which is being connected. There is great potential here in approaches to animals' geographies that place the nonhuman organism at the center of their spatial inquiries to map the "beastly places" and wider spatialities inhabited and performed by certain organisms.[82] Such an approach offers a dynamic understanding of place configured by the intersection of multiple processes with frequently discordant or incommensurable rhythms.[83] As I explore in chapter 8, such a topological understanding has the potential to map the "multiplicity" of biogeographies that overlap, intersect, or pass by in any ecology.[84] A swallow inhabits a world very different from that of a woodlouse; a captive elephant tramps an ecology different from that of its free-ranging kin. Meanwhile, different social groups associate with and enact wildlife in multiple ways.

DIFFERENCE

Taken together, these four concerns with hybridity, agency, immanence, and topology help to flesh out the character of a multinatural ontology of wildlife. This figures a lively world inhabited by diverse agents and propelled by many, frequently discordant processes. It has different rhythms, territories, and forms of (dis)connection. It differs markedly from the prevalent ontology of Nature outlined in the introduction. At the heart of this approach is a concern for difference.

This concept is central to conservation biology, a discipline frequently dubbed "the biology of numbers and difference,"[85] and to the forms of biodiversity conservation it informs. As I illustrate in the following chapters, an ontology of wildlife suggests a conception of difference alternative to that which has come to characterize much biodiversity conservation. Alongside its concerns for purity, conservation has tended to focus on extant diversity—the forms that make up the world at any given time.

There are good practical and political reasons for this approach, but it drains the life from any ecology, rendering the present eternal at the expense of the generative processes that keep any ecology alive. To address this problem, it is useful to differentiate between difference and diversity. Deleuze argues that "difference is not diversity. Diversity is given. Difference is that by which the given is given."[86] Difference here refers to what Deleuze terms the "virtual," the immanent potential within any assemblage to become otherwise. Understood this way, the focus shifts from the diversity of essential, existing beings to the processes of becoming. Becomings happen within organisms, between organisms, including at the interface of people and wildlife. As I argue in the chapters that follow, a focus on difference (rather than diversity) takes conservation biology back to its Darwinian roots[87] and resonates with new thinking on resilience and transformation, novel ecosystems, and the potential of reorganizing the spaces and times for conservation around connectivity.

This shift to difference does not solve a second problem, which relates to the ontological politics associated with choreographing conservation around different incommensurable units. An ontology of wildlife offers multiple forms of difference. In practice it suggests multiple post-Natural ontologies. Different ways of figuring the life that should be secured and let flourish imply very different modes of biopolitics. Targeting genes, individuals, species, cultures, populations, ecologies, processes, ecosystem services, and so on generates different politics and ecologies. The ontological politics associated with this multiplicity flags the degree to which any management decision is a biopolitical act.[88] Nature does not provide the answers. The nonhuman world shapes the possible outcomes, generally in ways beyond current human understanding and control. But the key decisions that concern this book are political. A commitment to difference does not render biodiversity conservation a contradiction in terms or imply the laissez-faire relaxation of all forms of human control. This would be an abnegation of the responsibilities posed by the Anthropocene. Instead, it demands better science, better politics, and new forms of human–environment relation. There is wildlife in the world that still needs conserving.

· 2 ·

NONHUMAN CHARISMA

Counting Corncrakes and Learning to Be Affected
in Multispecies Worlds

It is the summer of 2003, and I am on a small island in the Hebrides—a sparsely populated archipelago off the northwest coast of Scotland. It is a dark and windy night, and I am following Craig,[1] a conservation biologist employed by a large UK NGO. We are trying to count corncrakes (*Crex crex*), a light-brown bird about the same size as a pigeon (Figure 2), as part of a national census. This has not been easy. By day corncrakes tend to skulk in the long grass and are invisible. Their distinguishing feature (to humans) is the nocturnal call of a male bird in search of a mate—the loud, persistent, and metronomic "crex crex" from which the species gets its onomatopoeic binomial.

Over the summer Craig makes a series of nocturnal forays to listen for calls across his allocated territory. This is a large and inaccessible area with few roads. He walks and drives for miles, listening hard above the persistent wind to differentiate crakes from the similar sound of passing fence posts or grass on boots. Sometimes, he encounters multiple crakes. One calling male prompts others establishing territory to call in unison. He must disentangle and individuate this cacophony by detecting differences in rhythm, tone, and modulation. Through training and experience, he has learned to tune into the birds' acoustic ecology, calibrating his body to "learn to be affected" by the corncrakes' calls.[2] Each call must then be located and mapped, triangulating from several positions in the dark. Inventive males use local topography to project and amplify their craking. This

FIGURE 2. A corn-
crake amid typical
vegetation (*Crex crex*).
Photograph by Sergey
Yeliseev, Wikimedia
Commons.

complicates location. Mapping corncrakes requires an intimate, haptic knowledge of the local landscape.[3]

In the daytime I join Craig in bright sunshine on the edge of a farmer's field, which is being cut for silage. He is holding a radio tracker, a device that looks like a large TV aerial. It is emitting a series of beeps, detecting the radio signal given out by small tags that have been attached to corncrakes in the field. Tuning into the corn-crakes' new daytime frequency and calibrating between ear, eye, and his moving body, he follows the birds as they try to evade the mowing machine and lurking predators. He is interested in their behavior. Hunches are formed, observations are scribbled, and dismembered bodies are counted.

Counting corncrakes is a technological practice. It happens within an assemblage of skilled bodies, instruments, and recording devices. Counting corncrake is also passionate. Craig enjoys his crake-filled summer nights on the islands. He has been counting for a decade

FIGURE 3. A sunny day in the Hebrides. Photograph by the author.

and is filled with hope and wonder at every new arrival. He loves the intellectual satisfaction of counting, plotting, and tracking. Night and daytime hunting keeps him fit and can be exhilarating. Like many involved in UK conservation, he often volunteers. The promise of fun, wonder, and hope constitute the affective energies that motivate him to get involved.

The corncrake is a rare migratory bird that has been prioritized for nature conservation in the United Kingdom. I say much more about the science and politics of its conservation in chapter 4. In this chapter I dwell on these passionate interactions between Craig and the corncakes in the fields of the Hebrides. I take these as illustrative of the multitude of embodied and affective encounters between skilled humans, other species, and landscapes that underpin conservation. It is through these encounters that the flux of wildlife that I outline in the previous chapter comes to be sensed and known by conservation biologists. It is from these encounters that they develop and test forms of environmental management. It is on the authority of these interactions that Nature can be invoked. As I discuss in greater detail in the chapter that follows, it is ultimately from such encounters that wildlife gets framed and governed as biodiversity. It is here that the ontological choreography of conservation that I outline in the introduction begins.

I also want to focus on these encounters because they are done poor service by the official accounts of field science offered by conservation biologists. These scientific papers, textbooks, field guides, and management action plans struggle to account for the diverse agencies of habituated, skilled bodies. They downplay the influence of target species, living landscapes, and technical instruments in the generation of scientific knowledge. They are also wary of discussing the passions that power conservation.[4] These actors, skills, and affective energies seem taboo, their public acknowledgment threatening to undermine the credibility of the objective natural knowledge they have helped generate.

The public presentation of conservation is wedded to a positivist epistemology of Natural Science that promises a single route to Nature. As I explain in the previous chapter, this is tied to a dualistic understanding of the Human as a rational, disembodied "brain-in-a-vat"[5] revealing an objective and panoptic Nature seemingly unassisted by technology. This is a shame. First, because it provides an inaccurate and exclusive account of science in action that forecloses on a measured appraisal of conservation's current and potential partialities and negates a fruitful discussion of its politics. Second, because it conforms poorly to the private sentiments of many conservationists, who, like Craig, love wildlife in different ways and care deeply for its future. In my experience they are propelled by their passions, by the value of encounters, not by the instrumental logics of ecosystem services.

In this chapter I begin to articulate an alternative epistemology for conservation after the Anthropocene. An epistemology is a theory of knowledge. It establishes the procedures by which legitimate knowledge might be produced and thus what might be accepted as truthful. I build this account from the challenge to the modern figure of the Human that I introduce in the previous chapter and develop a more-than-human epistemology of conservation that acknowledges the diverse agencies of skills, bodies, and nonhumans whose absence I have identified. Remaining consistent with the ontological commitments of the preceding chapter, I aim to present conservation research as a set of tentative, multispecies processes of "learning to be affected" by the world. This is a multinatural model that acknowledges a multiplicity of natural knowledges. Here, natures are materially multiple and can be legitimately known in many ways. In this chapter I am particularly interested in the "affective logics" that guide the science of conservation.[6] An affective logic describes a particular mode of engaging with, knowing, and feeling toward wildlife.

To develop this analysis, I draw on selected elements of a rich body of work concerned with embodiment, skill, affect, and performance in the social sciences. This work offers a "more-than-representational"[7] account of human perception, knowledge, and subjectivity. This builds from long-standing and critical poststructuralist interest in the processes of representation to explore how technologies, bodies, and other materials come to shape human sense making and behavior, often in advance of cognitive thought. This work flags the significance of affect as a set of energies that flow between bodies, emerging from embodied encounters and adhering in particular places and landscapes. I narrate my engagement with these literatures in more detail in the sections that follow, which are organized around a discussion of the concept of nonhuman charisma.

NONHUMAN CHARISMA

I take the term *charisma* from conservationists who use the word *charismatic* to describe a set of species that have popular appeal. This is a positive accolade generally associated with "flagship species" that can be made to circulate in media and markets; the Asian elephant is a classic example. Conservationists invoke nonhuman charisma vaguely and generally only through the adjective *charismatic*. There has been limited effort to specify the properties and operations of nonhuman charisma and to differentiate its affective force.[8] In this chapter I understand nonhuman charisma to describe the features of a particular organism or ecological process that configure its perception and subsequent evaluation. Here, I am concerned with the charisma of nonhumans in the context of proximal encounters with conservationists—like Craig and his corncrakes.

Conservationists' discussions of charismatic species tend to focus rather narrowly on what I term *aesthetic charisma*. This describes the visual appearance of a species in print, on film, or in the spectacular encounters of ecotourism. I widen the scope of what we understand to constitute and configure charisma in order to encompass the material properties of an organism, which I term its *ecological charisma,* and the feelings engendered in proximal, multisensory encounters between a conservationist and their target organism. I refer to this as *corporeal charisma.* Together, these dimensions offer a three-part typology of nonhuman charisma, which I present in more detail in the following sections.

This account of charisma is relational and in keeping with the ontological commitments outlined in the previous chapter. Here, charisma emerges in relation to the parameters of different human bodies that are technologically enabled but still corporeally constrained. The ecological charisma of an organism is largely consistent across all social groups. In contrast, aesthetic and corporeal charisma vary greatly according to the context of the encounter and the ends to which the target organism is to be put. As I illustrate in more detail in chapters 6 and 7, nonhuman charisma has cultures, histories, and geographies. It is not determinate, nor is it easily manufactured. Farmers, hunters, and conservationists differ markedly in their relations and thus evaluations of different species. Charisma is contested. Hebridean farmers disagree with conservationists about the joy of nocturnal corncrakes calling outside their windows, to give an easy illustration. Charisma is also a more-than-human phenomenon. Asian elephants and corncrakes have their own ways of perceiving and evaluating the charisma of those with whom they cohabit. It is reasonable to expect that the charisma of corncrakes would be different (perhaps even nonexistent) to an Asian elephant, in the unlikely event that they encountered one another.

Ecological Charisma

Ecological charisma describes the anatomical, geographical, and temporal properties of an organism that configure the ease with which it is perceived by a human subject in possession of all their senses and with limited technological assistance. To explore the ecological charisma of different organisms, we can take an ethological perspective on human–nonhuman–environment interactions. Ethology is the well-established science of animal behavior. Strands of this work have recently received interest from more-than-human theorists engaged in "multispecies ethnography."[9] This is due in part to the rediscovery of the work of Jakob von Uexkull, an early nineteenth-century Estonian biologist and founding figure of ethology. Von Uexkull starts from an understanding of a being, human or otherwise, as an ecological entity immersed in its environment and performing a number of core behavioral characteristics. He terms these the "affects" of an organism. Together, an organism's affects determine its *Umwelt,* or way of being a subject in its environment. Through his famous example of the tick, whose world is structured around three primary affects, von Uexkull eloquently shows how the intersections of the *Umwelten* of different organisms determine their possible interactions within an ecological complex.[10]

Clearly, the human *Umwelt* is more complex and variable than that of a tick, but an ethological perspective on human–environment interactions foregrounds the common properties of human bodies that, in Katherine Hayles's terms, frame the "cusp" through which we make sense of the world.[11] These can be understood as our primary affects. All humans are warm-blooded, (potentially) omnivorous mammals. Most humans are bipedal, between 1.4 and 1.9 meters tall, terrestrial, and diurnal (rather than nocturnal). Unlike most terrestrial mammals that communicate with pheromones, we depend on vision and privilege visual knowledge—like birds and butterflies. We are in possession of five senses, but the type of acoustic skill that Craig embodies is rare. Human sensory organs make use of small portions of the electromagnetic, acoustic, and olfactory spectra for perception and communication.

With this ethological understanding, we can see how the physiological and phenomenological configuration of the human body puts in place a range of filtering mechanisms that disproportionately endow certain nonhumans with ecological charisma. The human *Umwelt* intersects more or less easily with those of other organisms. These intersections determine the detectability of an organism and the ease with which an interested human is able to tune in to its behavior. Detectability is influenced by a range of parameters, including size, color, shape, and degree and speed of movement. It also relates to aural characteristics such as the presence or absence of a noise, call, or song and the frequency and magnitude of this sound. Taken together, these constitute what naturalists call an organism's *jizz*.[12] *Jizz* refers to the unique combination of properties of an organism that allows its ready identification and differentiation from others. The nature and frequency of any human–nonhuman encounter also relates to the intersections between the space–time rhythms of the two organisms. Seasonality, migration patterns, diurnal ecology, and distribution on and in land, air, or water all shape whether and how often they encounter people. The jizz of an organism and the concurrence of its ecological rhythms with those of humans configure a species' ecological charisma. The parameters of ecological charisma would be very different if we had evolved with gills or night vision.

The corncrake possesses a reasonable degree of ecological charisma. In spite of its elusive diurnal behavior, the male bird is easily distinguished by its nocturnal call. By day male and female birds have a distinct skulking behavior and awkward flight that are easily differentiated with the naked eye. These aural and visual signatures are the most readily detectable affects of the bird that gives it its jizz. Asian elephants possess greater ecological charisma. They are large, distinct, and easily individuated. Like the corncrake, though, they are often nocturnal and, in spite of their size, still prove challenging to count. Elephant population totals are surprisingly inaccurate.[13] The vast majority of life-forms are not ecologically charismatic. Leaving aside bacteria, most eukaryotes (animals, plants, fungi) are small, similar, and inaccessible to people. They have not been named, researched, or monitored. As I explain in the next chapter, this has important consequences for the scope of conservation.

Von Uexkull's theory of *Umwelten* is multinatural, in the sense that he acknowledges a multiplicity of more-than-human subjects with different perceptions of the environment. He was a conservative figure, however, who was both opposed to Darwinism and enthusiastic about National Socialism.[14] His ethology is antipathetic to change and discord and gives little scope to the agency of technologies. It imagines a romantic, premodern world of organisms following the timeless harmonies of a creator.[15] These elements of his ontology are at odds with the approach to wildlife I outline in the previous chapter. Deleuze and Guattari have reworked von Uexkull's theory of affect for a world in which organisms (and other materials) move in space and time according to discordant rhythms.[16] Their approach to ethology has informed more-than-human thinkers like Vincianne Despret and Donna Haraway, who offer a hybrid and more open-ended model of human–nonhuman interaction conceived as processes of "learning to be affected" by the world.[17] Here, humans and nonhumans—farmers and elephants, corncrakes and conservationists—become what they are through situated interactions over time. Organisms display a degree of what biologists refer to as "behavioral plasticity," or what social scientists call culture.

This commitment to becoming and the processes of learning to be affected draws attention to the role of habit and embodied expertise in field science. Detecting organisms like the corncrake takes time, training, and skill. It involves cultivating dispositions that attune a listening body to the landscape. These generate multisensory familiarities that are often unconscious and difficult to articulate. They happen in advance of thought and are thus more-than-representational. Evoking the language of Deleuze and Guattari, Craig talked about his research as various acts of becoming. Tracking corncrakes involved "becoming-predator," striving for a sensory affinity with a fox or cat to tune in to the bird's ecology. Anthropologists and philosophers informed by Deleuze have written of nonmodern people becoming-animal through such time-deepened interactions.[18] Although they might be appalled at the comparison, skillful field scientists have much in common with Inuit herders or Amazonia hunters. The importance of these kinesthetic knowledges became clear to me during my time with Craig. Although I had read the field guides to the birds of the Hebrides and understood the corncrake census methodology, I arrived on the islands with no past experience. These texts were poor surrogates.[19] In spite of Craig's patient pedagogy, creaks became crakes; multibird cacophonies could not be disentangled; and on daytime inspection my corncrake maps had birds nesting in the sea.

An embodied perspective on corncrake surveillance and research also flags the significance of technology for the generation of natural knowledge. Once they become familiar, instruments like Craig's radio tracker act as sensory prostheses. They expand the scope of the human *Umwelt* and help overcome many of the perceptual constraints I have outlined. Submersibles, microscopes, thermal imaging, and tagging technologies have all dramatically expanded the spaces and range of life-forms that scientists can perceive. Heterodynes (radio receivers) and pheromone detectors help attune to unfamiliar nonvisual forms of communication, while time-lapse photography and paleoecological methods witness nonhuman rhythms occurring at speeds both faster and slower than the naked eye can behold. These developments have been revolutionary, but they have occurred relatively recently and are often expensive and unwieldy to use in the field.[20] The possibilities of field science are still strongly shaped by the ecological charisma of nonhuman organisms, which lays the foundations for further types of charisma.

Aesthetic and Corporeal Charisma

Aesthetic and corporeal charisma describe the properties of organisms that generate emotional responses among people encountering them. Aesthetic charisma relates primarily to encounters with visual media or certain spectacular modes of ecotourism. Corporeal charisma is concerned with feelings generated in proximal encounters in the field. The two are closely entwined. This division is not intended to replicate a mind–body division or to make a false distinction between representation and an authentic reality. The separation relates more to where the human–nonhuman encounters take place than to any clear qualitative difference between the modes of charisma they describe. Both aesthetic and corporeal charisma involve embodied practice and draw to differing degrees on the full panoply of senses. As I explain in more detail in chapters 6 and 7, mediated encounters with aesthetic organisms play an important role in triggering and affirming lived experiences in the field.

To understand the character and significance of these feelings, we need to explore a further dimension of the concept of affect. Here, affect can be understood to extend beyond material properties and habits to encompass the feelings, moods, and emotions experienced in embodied encounters. Social science has recently undergone something of an affective turn, and there is now an extensive and differentiated set of literatures upon which we could draw to engage with these feelings.[21] In the sections that follow, I draw from several different strands of this work, ranging from poststructuralist philosophy, psychoanalysis, and elements of evolutionary psychology. Close readers of these literatures may find inconsistencies in my account, but my aim here is to develop new concepts.

In mapping aesthetic and corporeal charisma, I offer a relational account of affect in which shared structures of feeling bubble up within particular constellations of people, technologies, and other nonhumans. These have differing durations, from singular events to persistent and embedded attachments, anxieties, and affections. Here, affect can be governed, but it is not determined—by either evolutionary imperatives or universal human and other natures. These types of charisma thus have distinct cultures, but also notable consistencies. As such it is possible to identify particular *affective logics* that guide how people act in relation to particular species and landscapes. Reason or rationality, so central to the public presentation of modern science, is not the absence of affect but a particular affective logic. It is also a rare thing.

As we saw with Craig in the Hebrides, familiar habits and the promise and potential of particular feelings provide a powerful influence on action. The volunteer or underpaid conservationists with whom I have worked are drawn to their field by reasons that escape the narrow confines of instrumental rationality. These men and women are not in pursuit of personal profit or the future cure for cancer. They do not value corncrakes or elephants as deliverers of ecosystem services. Instead, such species offer them a range of powerful and desirable affective encounters, from specific moments of joy and despair to slow-burning aspirations and fears. As I examine in chapter 7, species like the elephant (and to a lesser extent the corncrake) have vital economic and political power in the global assemblages of conservation by virtue of their popular appeal; they offer a platform for evoking affect and mobilizing important resources. The end result is not perfect, nor is it reasonable. One of the central arguments of this book is that the character of such affective logics shapes the cultures, practices, and, ultimately, the politics and economics of conservation. As I explain in more detail in the following chapter, affect matters across all the arenas through which conservation proceeds.

Aesthetic Charisma

Aesthetic charisma is the type of charisma most familiar to discussions among conservationists. It describes the visual properties of certain organisms that would normally be described and presented as charismatic in marketing and advocacy materials. In conservation parlance this charisma is generally seen as a positive accolade associated with organisms to which one could append the adjectives *cute* and *cuddly* (the panda) or *fierce* and *deadly* (the tiger). As a result of the increased importance attached to conservation marketing and a growth in the resources attached to advertising and campaigning, there has been a steady sophistication and diversification of the aesthetics of publicity materials. A wider range of affective logics is now deployed, which I discuss in more detail in chapters 6 and 7.

From the outset we should be clear that, in my understanding of it, aesthetic charisma does not guarantee a universal positive response from human publics. The landscape of aesthetics can be starkly polarized in relation to both the anatomical character of popular species and the feelings they invoke in different audiences. There is a consistent vital force at work here that seems to bring some species to the forefront of popular attention, but the responses they engender are underdetermined. For example, both elephants and cockroaches are charismatic, but their charisma can engender strong and divergent responses. Different organisms can be both awe-some and awe-full.

Popular responses to the aesthetics of organisms appear to be arranged along an axis of anthropomorphism, which has been the subject of numerous and diverse theories. For example, the ethologist Konrad Lorenz argues for the existence of instinctive human preferences toward organisms that exhibit some combination of a big head, upright posture, flat face, round profile, feet-like hands, large eyes, and soft fur: in other words, organisms that look like human babies.[22] Lorenz echoes much earlier work by Charles Darwin on the similarities in expressions between man and animals.[23] Although Darwin does not explicitly discuss the effect of animals on humans, he draws attention to homologous expressions that cross the human–animal divide. E. O. Wilson and other authors have offered sociobiological explanations, arguing for the existence of universal forms of "biophilia" and "biophobia" triggered by particular risky species (such as snakes and spiders) or desirable landscapes (parkland).[24] They claim a genetic signature for these affections and argue that avoiding or dwelling in such landscapes would have conferred evolutionary advantages on our ancestors.

Social scientists have sought cultural explanations of anthropomorphism, identifying the importance of nonhuman resemblances to key parts of the human anatomy that have been taken as signifiers of the human. For example, Owain Jones draws on and extends Emmanuel Levinas's concept of the "face" to explain anthropomorphism.[25] Levinas argues that the face—and more specifically the eye—is the vital medium through which all human communication interaction occurs. By extension nonhumans with faces and a bifocal gaze will attract attention. In a similar, though anatomically different, maneuver, Martin Heidegger emphasizes the "hand" as the human signifier.[26] Work advocating for animal rights has mapped and promoted interspecies affection as a result of the expression of shared emotions—principally suffering but also play, love, and grief.[27]

Kay Milton explains how this form of "human extensionism" in environmental ethics operates through the allocation of "personhood" to particular individual nonhumans.[28] These anthropomorphic criteria for aesthetic charisma operate most powerfully in the case of animals that display a form of reciprocity to human action and concern. I explore how this manifestation of aesthetic charisma is evoked in moving images of elephants in chapter 6. In chapter 7, I offer a more general discussion of how this "cuddly charisma" motivates a popular desire for touching encounters with captive animals. These are emotional encounters in which touch plays a central role. This practice sits as the sometimes problematic interface of concerns for conservation and animal welfare. Demand for touching encounters has created a market for captive animals with troublesome implications for individual welfare and free-ranging populations.

Aesthetic charisma also stimulates a range of negative attachments. Certain organisms and, indeed, whole taxa may trigger strong and visceral feelings of disgust and even panic among those humans they encounter. The social manifestations of various forms of entomophobia (insects), ophidiophobia (snakes), and murophobia (rodents) are well documented though, of course, far from universal.[29] There is little existing work that seeks to explain the causes of this negative aesthetic charisma or explore its consequences. One interesting explanation comes from the psychologist James Hillman. Concentrating on insects, Hillman analyses a psychological condition that he had identified in a number of his patients, which he terms "going bugs." He identifies four characteristics that, he argues, provoke the "frightening fantasies" that many people experience (often only weakly) when they encounter insects. He terms these *multiplicity, monstrosity, autonomy,* and *parasitism.*[30]

Hillman argues that the sheer multiplicity of insects, such as ants and flies, "numerically threatens the individualized fantasy of unique and unitary human beings."[31] Many people find it difficult to understand taxa where the individual is so radically subsumed by the many, where the subject is unimportant and an individual nonhuman cannot be personified. As Deleuze and Guattari point out, organisms that swarm in packs threaten the modern understanding of the bounded subject.[32] Monstrosity refers to the otherness of much insect physiology and behavior, which poses "body-space challenges" to the anthropomorphic anatomical norms I identify.[33] This alterity is evoked and amplified by the popular incarnation of aliens as insects in cartoons and film.[34]

In contrast to animals that show reciprocity, insects are understood to be more autonomous; they rarely react to human presence. Many insects communicate by pheromones and seem to resist all forms of human control and domestication. As Hillman puts it, "You can charm a snake, supposedly, or rub the belly of an alligator . . . but it is pretty damn hard to get a bug to do anything you want."[35] Finally, Hillman explains that "not only do bugs invade your realm, they also live off your property and share your body, thriving on your vegetative roots and pet flesh."[36] Insects as parasites transgress modern moral geographies that mark out the spaces and practices of bodily hygiene, domesticity, and civilization.[37] In identifying these four characteristics, Hillman echoes the importance that Lorenz places on anthropomorphism. In the case of insects, it is their radical alterity to humans in terms of size, ecology, physiology, aesthetics, and modes of social organization that engenders popular feelings of antipathy and distrust. In many ways, insects have the characteristics of what Julia Kristeva terms the "abject"—the breakdown of meaning that results from being confronted and overwhelmed by an other.[38] This alterity is not always interpreted as a negative accolade. Many scientists—not least entomologists—admire and respect those organisms whose bodies and social worlds are far from a human norm.[39] In contrast to the human extensionism of an anthropomorphic affection for cuddly charisma, which leads to a sense of sympathy for a fellow subhuman, this interest in alterity is grounded in a sense of respect for the other in all its complexity, autonomy, and monstrosity. For E. O. Wilson writing about ants or Dave Goulson on bees, it is this very difference and what it tells us about ecological and evolutionary processes that make insects interesting.[40] There is an affective logic of curiosity at work here, which I expand upon in the following section. Although they welcome the profile and resources generated by the spectacular aesthetic and/or cuddly charisma of flagship species, some scientists and conservationists link the prevalence of anthropomorphism to the neglect of the vast majority of unaesthetic life. Here, *anthropomorphic* becomes a term of abuse. In its most extreme forms, this results in what Steve Baker has termed "anthropomorphobia"—a fear of anthropomorphism. Here, an aversion to sentimentality drives a dogmatic defense of objectivity or a romantic (and frequently masculine) attachment to a purified wild and a concomitant antipathy to the domestic sphere.[41]

To summarize, aesthetic charisma refers to the distinguishing properties of an organism's visual appearance that trigger affective responses in those humans it encounters. Aesthetic charisma requires ecological charisma but is not determined by it. Instead, it is strongly related to the degree of alterity of an organism in relation to a broad set of anthropomorphic anatomical and behavioral norms. The corncrake is not

especially charismatic in this regard. Although it has come to feature in
the tourism and conservation marketing materials for the Hebrides, its
squat, tan, avian body and elusive, autonomous ecology do not seem to
evoke strong popular affection.

Corporeal Charisma

This final type of charisma refers to the feelings engendered by different organisms in proximal, practical interactions with humans. This is
a diverse form of charisma with many possible manifestations. Here,
different groups of organisms come to be associated with and valued
within distinctive affective logics that shape how interested human
subjects come to relate with them. These range in quality, intensity,
and duration and are informed by the ends to which the organism is
to be put. They are shaped in part by the ecological charisma of the
organism, as well as its other material properties. For example, farming has distinctive affective logics. It involves a range of habituated
practices, technologies, and domesticated plants and animals. This
practice values plants and animals that can be subsumed to a logic
of (re)production—that are useful, edible, resilient, amenable, and
(in much of the modern world) profitable. In contrast, sport hunting
involves an affective logic that values animals that are large, autonomous, intelligent, detectable, and ultimately killable.

Farming and hunting are beyond the scope of this book. I include
them here to indicate the diversity of ways in which an organism can
become charismatic. Here, I am most concerned with the affective
logics of conservation, but even this fairly narrow practice involves
a range of different ways of engaging with wildlife. In this section I
illustrate corporeal charisma through a discussion of two common
(though still fairly esoteric) affective logics that characterize the practices of field scientists like Craig. I have termed these *epiphany* and
jouissance. In chapter 7, I examine the affective logics of spectacle,
touch, and adventure associated with popular practices of ecotourism.
I explain how the commodification of this form of charisma generates
lively capital that increasingly shapes the scope, conduct, and politics
of conservation. I compare affirmative encounters with the experiences of those (frequently marginal publics) exposed to the dangers of

living close to charismatic large animals. An affective mode of engagement is not necessarily positive, for either humans or other species.

Epiphany

Epiphany describes the common autobiographical reference made by many conservationists to a specific transformative event involving an intense encounter with a particular organism, often in a notable location. Usually, these took place in childhood and have been made sensible through retrospective narration as shaping subsequent professional or voluntary practice. These epiphanies inscribe a memory and plant a seed that becomes a lifetime attachment, interest, and concern. Sometimes, they refer to just one event, such as seeing a rare bird. Others concern repeated or seasonal encounters: a sequence of events such as migration, a tree shedding its leaves, or regular trips to a local nature reserve.

These epiphanies are visceral and emotional but also very difficult to articulate. The cultural historian and enthusiastic birder Mark Cocker provides a compelling evocation in recounting an early birding experience:

> Then someone spotted an odd bird and it was instantly apparent I'd never seen one before. It was about the size of a curlew, yet not the same anonymous grey-brown color and with an indefinable quality of beauty and strangeness. It floated away across the moor and then suddenly wheeled around and turned towards us, its silent and loosely bowed wings knitting a course through the up draughts in long exaggerated beats, not unlike a giant bat. . . . It was a Short-eared owl, a bird in aerial display asserting its breeding territory with that fantastic see-saw action. . . . This bird was the first I'd ever seen, I recall, in fact, it was my ninety-ninth species, and it was wonderful. Before that moment I had, like every young keen birder, compensated for experiences of the real thing with long hours poring over bird books and bird pictures. But on Goldsnitch Moss I realized, perhaps for the first time, by how much life can exceed imagination. A Short-eared owl

had entered my life and for those moments, as it swallowed me up with its piercing eyes, I had entered the life of an owl. It was a perfect consummation.[42]

For Cocker the wheeling, diving bird, at one with the same wind that buffets his body, provides an organism model to which he aspires in his birding practices: the bird provides a line of flight from his earth-bound, heavy identity as a clumsy human toward a new lighter mode of being. In entering the life of the owl and becoming consumed by it, Cocker is momentarily carried away. On his return he looks at things in a different way. These moments of becoming and consummation have many of the qualities of the enchantment identified by Jane Bennett.[43] They frequently animate the best popular natural history writing, though never the papers of conservation biology journals. Moments of connection like these give amateur and professional naturalists such pleasure in their work. They are addictive. Craig fondly remembered his first childhood encounter with a corncrake on holiday in the Hebrides. This memory is affirmed annually in his first springtime encounter with a returning corncrake. The promise of this moment helps keep him going through a long, dark winter.

Jouissance

I use *jouissance* to describe the intense and sometimes disconcerting feelings of intellectual satisfaction experienced by self-described scientists in their everyday knowledge practices. I take the term from Julia Kristeva, for whom it refers to the pleasure experienced in the presence of meaning.[44] Jouissance can manifest itself in innumerable ways and is offered up in an unequal fashion by different organisms. One common example is the joy of identification and the making and completing of species lists—esoteric practices at the heart of amateur and professional field science. Birders and other naturalists love lists; they develop day lists, life lists, and patch lists that give direction, territory, temporality, and status to their professional and leisure activities. Field naturalists pride themselves on their ability to identify species. This is an incredibly skillful activity and takes years of training and experience. Once one is familiar, or striving to become so,

identification offers a number of what Cocker terms "seductive plea-
sures."[45] These include the quiet sense of satisfaction that comes when
the components of the world fit the units and schema with which you
are familiar. Identification offers a sense of intellectual ability and
completeness.

Cocker partly attributes the disproportionate popularity of birds
in the United Kingdom as subjects for population surveillance to the
manageable size of the list of resident species. There are about only
200 to 250 birds in the United Kingdom that an enthusiast would
be likely to encounter on a regular basis, in contrast to nearly 7,000
diptera (flies)—few of which are as easy to identify and differentiate.
British birds are an accessible and satisfying group.[46] An individual
collecting and listing sightings is likely, with a bit of effort, to see
most of them in a lifetime. Furthermore, the length of the bird list is
of a suitable size to allow competition among "twitchers," who collect
a range of different lists of birds spotted. In relation to the jouissance
of list making, species that are genetically or individually rare are disproportionately endowed with this corporeal charisma, as are taxa that
are easily differentiated from their taxonomic kin and neatly divided
into coherent species—such as beetles or butterflies.

For other scientists concerned more with ecological dynamics
and processes, there is a quiet, fragile satisfaction in disentangling
the mysteries of the nonhuman world through patient investigation.[47]
For example, Craig spoke with pride about his work mapping the
ecology and dynamics of the corncrake population, the sense of ac-
complishment it had delivered that was accentuated by the effort and
frustrations he remembered from his fieldwork. In articulating the affective logic at work in his (and his colleagues') laboratory research, he
talked about "becoming-calculator," in comparison to the becoming-
predator he had identified in his pursuit of corncrake in the field. This
involved disciplining his body, suppressing its disruptive and distract-
ing demands to enable cerebral contemplation, technical calculation,
and abstraction.[48]

His comments draw attention to the different affective logics of the
lab and the field, which I describe in more detail in chapters 4 and 5.
Here, the field is a public and contingent site in which one must be
open to new developments and the promise of a novelty and epiphany.

In contrast, in the lab one must become calculator, achieving a degree of simplification, control, and abstraction to produce general knowledge. As I show in chapter 6, for scientists seeking to enroll volunteers to assist (or at least fund) their field science, there is a constant tension between the scientific demands for disciplined and patient observation and the tourist demands for epiphany and adventure. It is difficult to convince a fee-paying volunteer of the merits of an absence when that absence is the elephant they are desperate to encounter.

CHARISMA AND THE CONDUCT OF CONSERVATION

In this chapter I have begun to develop an alternative epistemology for wildlife conservation that helps avoid the problems that have been identified with positivist natural science and the modern figure of the rational Human Scientist. Here, knowledge about the nonhuman world emerges out of situated, embodied, and technological encounters with the nonhumans that are the subject of research. The bodies of scientists are vital for this endeavor. It is only through training and experience that a scientist can learn to be affected by their target organism, ecology, or process. Technologies enhance and extend the possibilities of perception and recording. Science is propelled and guided by scientists' affective attachments to particular species and places. Habits and passions matter here and should be acknowledged, cultivated, and celebrated. There are multiple affective logics at work in conservation that shape what knowledge gets produced and what is accepted as a legitimate account. This is a multinatural approach, with the potential for difference and discord. Finally, natural knowledge is shaped by the relational agencies or biopower of nonhumans—in this case, in the form of nonhuman charisma.

This embodied epistemology helps account for some important patterns and processes in the scope and operation of the biopolitics of contemporary conservation. I say more about these in the chapters that follow. By attending to the ecological, aesthetic, and corporeal charisma of different organisms, we can begin to explain the prevalent units and partial taxonomy of contemporary conservation. For biologists in the field, working with and informed by various lay knowledges, the species provides an intuitive unit. Discrete species—

especially those that can be identified by morphological/audible differences—provide accessible units for listing, counting, mapping, and auditing the success of conservation interventions. Species provide a handy index of current ecological composition for monitoring change. They are the canary in the coal mine. Habitats and diffuse and nonlinear ecological processes that characterize the ontology of wildlife outlined in the previous chapter are much more difficult to bound, define, and monitor.

The ecological charisma of particular organisms strongly configures the taxonomic scope of conservation. Factors such as size, shape, distribution, visibility, distinctiveness, and detectability all help account for the frequently observed partialities in species description, the extent of biological knowledge, and the recording and monitoring activity that characterize historic and contemporary science and conservation. These patterns can be compounded by the power of aesthetic and corporeal charisma in shaping scientific and popular enthusiasms, funding, and campaigning for research and conservation. As I explore in more detail in chapter 7, the growing dependence of in situ and ex situ conservation on market-based mechanisms ties the future of wildlife to the affective economies of ecotourism and NGO fund-raising. Here, valued encounters become commodified. Charisma generates a form of lively capital whose taxonomy has important and frequently fraught implications for nonhumans kept in captivity, wildlife in reserves, and those publics forced to cohabit with global charismatics. Acknowledging and celebrating affect is important, but it is not a simple solution for conservation after the Anthropocene.

· 3 ·
BIODIVERSITY AS BIOPOLITICS

Cutting Up Wildlife and Choreographing Conservation
in the United Kingdom

The term *biodiversity* was invented by a small group of conservation biologists in the mid-1980s.[1] This buzzword entered popular consciousness at the United Nations Earth Summit in Rio de Janeiro in 1992, where 155 states signed the Convention on Biological Diversity (CBD). Biodiversity promised a new way of understanding and governing the environment; its advocates sought to rationalize existing conservation and galvanize future action. The subsequent increase in the use of the term in scientific and policy circles has been meteoric. Biodiversity energized and has been institutionalized within the scientific discipline of conservation biology—a self-proclaimed "crisis discipline"[2]—whose adherents research and advocate scientific strategies for biodiversity conservation. In differing fashions biodiversity has been incorporated into the national environmental policies of signatory states. Biodiversity arrived in the United Kingdom, which is the principal focus of this chapter, in 1994, when the government published its Biodiversity Action Plan.[3] Biodiversity has become a "global nature": a hegemonic framework for conservation science and policy.[4]

The CBD defines biodiversity as "the variability among living organisms from all sources including, inter alia, terrestrial, marine and other aquatic ecosystems and the ecological complexes of which they are part; this includes diversity within species, between species and of ecosystems."[5] Conceived this way, the scope of biodiversity is panoptic; it encompasses everything, everywhere. It incorporates the three interwoven biological scales of genes, species, and ecosystems on earth, in water, and in the sky. In most cases this definition is further expanded to include the processes that link these scales and spaces.[6] In practice biodiversity did not emerge anew. It came out of and is thoroughly dependent on the embodied, affective, and technological encounters between multiple species that I describe in the previous chapter. Biodiversity happens in an assemblage. It inherits and is haunted by particular knowledges, habits, instruments, territories, and practices.

In this chapter I examine biodiversity as biopolitics. As I explain in the introduction, I understand biopolitics to describe late-modern ways of securing life at the scale of the population (or other aggregations of individuals). Biopolitics involves productive and destructive processes through which life is made to live or left to die. It involves interacting with diverse and lively forms of biopower: the agencies of organisms and ecologies whose lives are to be secured. There is a growing body of empirical work mapping different modes of nonhuman biopolitics, including agriculture, forestry, fishing, biosecurity, animal welfare, hunting, and pet keeping.[7] Each has its own aims, privileged knowledge practices, and desired norms and subjects. Cutting across these modes, this work identifies a range of common and significant biopolitical practices concerned with understanding and intervening into the character, distribution, and dynamics of nonhuman populations. These include knowledge practices for identifying, classifying, counting, surveying, mapping, and calculating. Databases and models are key here. These knowledge practices inform practical management actions like culling, fencing, translocating, vaccinating, breeding, and planting.

To date there has been relatively little work on conservation as biopolitics.[8] This is somewhat surprising, as in its recent concern for biodiversity conservation is an archetypal biopolitical practice. Biodiversity conservation seeks to secure the future health and diversity of life, which is understood as a vital and threatened yet unruly and unpredictable resource. Biodiversity is understood in aggregate terms, most commonly as dynamic populations of species that can be known, modeled, and governed through strategic interventions aimed at both human and nonhuman subjects. Biodiversity conservation is informed by a desire for panoptic knowledge, comprehensive accounting, and efficient, instrumental management. It seeks to rationalize existing practice through the development and dissemination of standardized criteria and modes of interacting. This involves extensive and diverse knowledge practices, material instruments, and practical, skilled interventions. Efforts to secure and enhance life inevitably involve letting other life die, especially at a time of accelerated extinction. My aim in this chapter and the two that follow is to begin to address this gap. I present biodiversity as a form of environmental governance actively shaping human and nonhuman subjects and the wider ecologies they inhabit. I seek to contextualize the encounters discussed in the previous chapter, to trace how they are political, in the broad sense that they are actively bringing new worlds into being and thus foreclosing on other possibilities. Here, I am interested in what sociologists of science have described as "ontological choreography" and "ontological politics" of science in practice.[9] I trace how biodiversity conservationists "cut up"[10] the flux of the wildlife to create the units, theories, models, and territories that come to inform practical action. I examine the derivation and performance of a framing of wildlife that departs from the panoptic aspirations of the official definition of biodiversity to focus on a charismatic set of species. In chapters 4 and 5, I critically evaluate the differences between two equally successful, but very different, ways in which the biopolitics of conservation takes place. The balance of my analysis is toward the nonhuman consequences of different ways of conducting conservation, but in the hybrid immanent ecologies of the Anthropocene, conservation inevitably impinges on the discordant values, practices, and livelihoods of diverse human actors. Here, biodiversity comes to inform contested environmentalities geared toward shaping good conservation subjects.

BIODIVERSITY IN THE UNITED KINGDOM

I focus my analysis on the United Kingdom and draw on research I conducted on the arrival and accommodation of biodiversity into its national conservation infrastructure.[11] The United Kingdom offers an interesting case study. It is the country (or more accurately, increasingly devolved and disunited set of nations) with arguably the oldest, largest, and best-funded infrastructure for biological research and conservation in the world. It has some of the most extensive national surveillance programs and associated biological databases. It hosts several of the oldest, largest, and most powerful conservation non-governmental organizations (NGOs). These organizations have wide public support and extensive land holdings and financial resources.[12] The United Kingdom has pioneered important elements of the science and practice of nature conservation and has been an active exporter of conservation from its colonial era to the present day.[13] Many of the characteristics of UK conservation that I outline are common elsewhere. British people are also famed, among other less desirable accolades, for being a nation of animal lovers.[14]

Where conservation in the United Kingdom (and in much of western Europe) differs from that in many other parts of the world is in its limited embrace of the processes of neoliberalism, whose influence critics have identified in nature conservation elsewhere and in other domains of environmental policy. This is changing, but during the period on which I focus (1992–2002), the political economy of conservation in the United Kingdom was dominated by a range of NGOs working in conjunction with sympathetic statutory authorities at the national and European scale. These groups were largely opposed to the logics of private property, markets, and commodification. Conservation management was funded through volunteer donations, direct public payments, and, most significantly, taxpayer-funded agro-environmental subsidies delivered through the EU Common Agricultural Policy.

Conservation research and monitoring was conducted largely by amateurs, in house at NGOs, or in public universities and research institutes. Although the United Kingdom has some of the most concentrated levels of private land ownership in the world,[15] the political economy of its conservation is characterized by volunteerism, donation, and subsidy. This political economy comes to frame the mode of biopolitics that I detail in this chapter and the two that follow. In chapter 7, I trace how the biopolitics of conservation changes when markets and commodities become more significant

At the time of the Earth Summit, nature conservation in the United Kingdom was well founded. It encompassed a wide diversity of organizations whose membership encompassed nearly 10 percent of the population. It enjoyed a large area of designated land and had a solid legal framework. There was a long history of scientific research and monitoring. There was, however, a great deal of organizational overlap, duplication, and confusion. Conservation was an amateur practice motivated by a wider range of intellectual, political, and ethical enthusiasms. Relevant data were scattered and often incommensurable. British conservationists were early adopters of the neologism *biodiversity*, recognizing its political and economic potential and the mandate it offered for organizational change. They pressurized the UK government to attend, ratify, and implement the CBD.

The result was the UK Biodiversity Action Plan (UKBAP), which was drawn up and swiftly published two years after the Earth Summit. Scientists and civil servants on the steering group responsible for drafting this report were charged with rationalizing existing practice to bring it in line with the panoptic ambition, instrumental logics, and normative discourse of the CBD. This document offers an ambitious blueprint for conservation. It summarizes the state of biological knowledge, sets priorities for future action, offers a model through which they might be delivered, and establishes targets by which its implementation might be audited.

The UKBAP and associated documentation are dry and technical. They have much more to say about the how of conservation than why we should conserve. Where reasons are offered, a utilitarian rationale is paramount. The text invokes the (then novel and fashionable) logics of sustainable development, arguing that "biodiversity should be treated as a global resource to be protected and conserved according to the principles of ecological, economic and social sustainability."[16] Biodiversity, they argue, can be evaluated and managed as a resource according to the economic criteria of rarity and threat. In contrast, the aesthetic and intrinsic values of biodiversity are downplayed, as they "cannot be readily quantified."[17] Normatively, this utilitarian ethic constitutes a universal, scientific appeal to people to look after themselves, their way of life, and their future dependents. Biodiversity conservation is here understood as the rational desire to protect our ecological life-support system and the present and future inputs to our economy.

The main policy outcome of the UKBAP was the identification

of 391 species whose populations would be subject to individual species action plans. A species action plan is a standardized document summarizing the current status of the species, the factors leading to its decline, and the action currently under way. The plan then outlines objectives, targets, and proposed actions for conservation.[18] These are a ubiquitous biopolitical technology in biodiversity conservation. For reasons I expand upon, the species has been taken largely as the principal ontological unit for biodiversity conservation (in the United Kingdom and elsewhere). In the UKBAP, delivering the 391 species' action plans became synonymous with securing the future of UK biodiversity. The processes by which these plans were derived and through which they were to be implemented thus constitute the foci of this inquiry into the biopolitics of UK biodiversity conservation.

Figure 4 provides a schematic visualization of biodiversity conservation in the United Kingdom. It identifies four main arenas through which wildlife passes in being framed and governed as a subset of species. The figure shows that in order for a species to be conserved it must first be described (Arena 1). Here, a collection of similar organisms are given a discrete taxonomic identity (ideally both a scientific binomial and a popular name), classified by a qualified taxonomist, and have a unique "type specimen" lodged in an accessible (preferably digitized) collection for subsequent cross-reference. Ideally, a species would feature in a field guide to enable subsequent identification. In order to understand the distribution and dynamics of a species' national population, it must be surveyed (Arena 2[a]). Here, organisms conforming to a species identifier are counted as a population over a suitable spatial and temporal scale according to consistent methodology to generate an accessible, standardized, and numerical dataset. Regular surveillance is vital for assessing the efficacy of conservation actions. A species must be researched to establish the causes of observed or potential population dynamics (Arena 2[b]). This requires interested, skilled, and resourced researchers; relevant published research and potential publishing outlets; research instruments; and accessible field sites.

To qualify for conservation action, a species' status must be evaluated (Arena 3[a]). Here, existing biological records and ecological research are collected, rendered commensurable, and aggregated. They

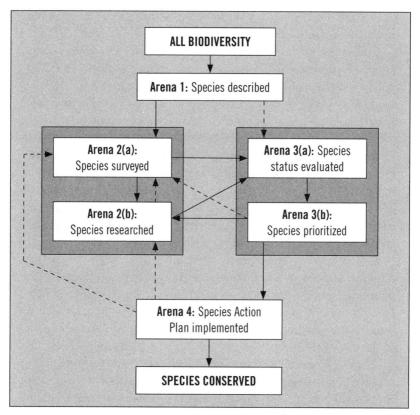

FIGURE 4. A schematic visualization of biodiversity conservation in the United Kingdom. A species' passage is not necessarily linear, though all species must first be described. Species with the longest history of natural history interest will proceed in order through arenas 1 to 4 (shown in bold). In contrast, other taxa (like many invertebrates) may not be researched and monitored until they have first been prioritized. As such, arenas 2 to 4 should be understood as interwoven, with different species tracing different trajectories between them (shown by the broken lines).

are evaluated in relation to international and national criteria of rarity and threat—the most common being those that inform the International Union for Conservation of Nature (IUCN) red list of globally threatened species.[19] The United Kingdom also has a national list of species of conservation concern. If a species is evaluated as threatened, it might then be prioritized for conservation (Arena 3[b]). This offers legal protection, creates a political mandate for action, and generates economic resources. Finally, a species must have an action plan successfully implemented on its behalf (Arena 4). This involves a wide range of interventions, depending on the species' ecological requirements and the nature of the threat. Common practices involve the acquisition and designation of land; legislation and subsidies to ensure sympathetic land management; and publicity, education, and advocacy to change human behavior.

The technologies, spaces, practices, and bodies that constitute these four arenas make up the material assemblage of UK biodiversity conservation. Many of these arenas are generic, and Figure 4 provides a useful model for understanding biodiversity conservation worldwide. Activities in each of the arenas cut up the diversity of life to generate manageable units, categories, models, maps, and other abstractions that come to guide practical action. It is through the diverse encounters that take place in these areas that these abstractions get performed and the biopolitics of UK conservation takes place. Here, I am particularly interested in how the diversity of wildlife introduced in chapter 1 gets framed and filtered through the operations of each of these arenas.

Figure 5 gives an illustration of this process. Here, the potentially infinitesimal diversity of UK wildlife is understood to comprise roughly 96,000 species. This is the total number that had been described when the UKBAP was drawn up. The population dynamics of about 13,000 of these had been surveyed,[20] and some 1,252 had had their status evaluated according to consistent criteria produced by the IUCN.[21] Of these, 391 were prioritized for action plans. Through this process the diversity of UK wildlife is reduced to 391 species. As I explain, there are distinct taxonomic partialities in the scope of each of these filters that relate to their dependence on the encounters I detail in the previous chapter. As a consequence the biopolitical assemblage illustrated in Figure 4 can be understood as an "oligopticon,"[22] an inevitable and necessary, but nonetheless partial, framing at odds with the panoptic aspirations of biodiversity.

Before zooming in on the activities and impact of UK biodiversity conservation to explore this filtering, it is important to contextualize

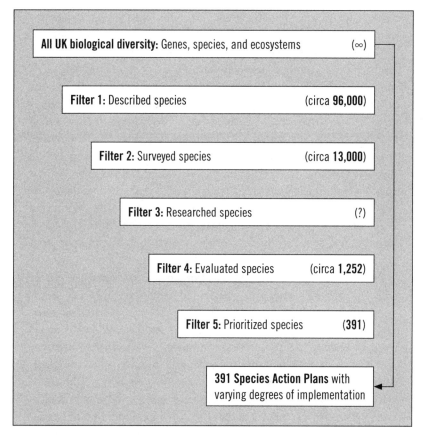

FIGURE 5. Illustration of the filtering mechanisms performed in the different arenas of the UKBAP.

and bound some of the claims to significance that follow. First, I should be clear that the in situ conservation of a prioritized species has beneficial (and sometimes detrimental) effects on the wide range of organisms with which it shares its habitat. Indeed, some of the species selected for the UKBAP were prioritized for their keystone or umbrella status, in the hope that their salvation would ensure the salvation of a range of valued others. As I explain in more detail in chapter 7, the ecological concept of a keystone species is different from that of a flagship species. These are iconic species with popular appeal, mobilized to build interest, support, and funds for conservation.

Second, although biodiversity might be hegemonic in conservation, it is one among many competing ways of managing the environment. In this competition it is certainly not all-powerful. In spite of a significant expansion in knowledge, territory, resources, and support since the Earth Summit, biodiversity is (even on its own terms) failing.[23] When the signatory governments gathered in Nagoya in 2012 to report on progress twenty years after the CBD targets, the story was of continued and sometimes accelerated declines. This is true in the United Kingdom as much as it is globally. Paradoxically, as biodiversity flourishes as a governance regime so the nonhuman difference it seeks to conserve continues to ebb. Agriculture, fishing, forestry, and other productive forms of environmental management continue to take their toll.

Finally, beyond the limited set of organisms nurtured by agriculture, the species that are faring best in contemporary hybrid ecologies are those most able to occupy its modified spaces and spatialities. The invasive, nonnative "global swarmers" and feral "synurbics" encountered in chapter 1 trouble conservationists as biosecurity threats—pest species that threaten biodiversity and circumvent human efforts toward their control. As I caution in the introduction, we should be wary of the popular anthropocentric metaphor of biodiversity conservation as an ark for the Anthropocene. The biopolitics of biodiversity will shape but not determine future ecologies. There are powerful inhuman natures at work here on a dynamic and warming planet that will shape future ecologies.

Figure 4 describes a wide range of multispecies relations too numerous and diverse for me to do justice in this chapter. In the following chapter, I trace the passage of the corncrake through these different arenas. Here, I focus on four significant knowledge practices and performative interventions within the wider operations of UK biodiversity conservation. These help illustrate the utility of conceiving of biodiversity as biopolitics and help account for the operations and scope of the UKBAP oligopticon. I start with the use of species as the primary ontological unit for biodiversity before describing species description, surveillance, and action plan implementation.

SPECIES: AN INTUITIVE ONTOLOGY FOR BIODIVERSITY

In my first ethnographic and textual encounters with conservation biology, it came as something of a surprise to discover that there is still such a great deal of uncertainty and debate within the discipline over the nature of the basic units that should be used to organize practical conservation. In theory, as previously explained, biodiversity encompasses the three scales of genes, species, and ecosystems. In practice it is difficult to identify and quantify all three of these in the field. Although there have been significant advancements with the advent of DNA bar coding, the technology for identifying genetic diversity is relatively new, expensive, time consuming, and practically inoperable for many species across a whole field site.[24] On a larger scale the clear delineation and classification of ecosystems is made problematic by their lack of spatial and temporal boundaries. Although the differences between a desert and a rain forest are clear in isolation, boundaries are often fuzzy and dynamic.

As a consequence, and given the imperative for action that motivates biodiversity conservation, the majority of taxonomists and conservation biologists have settled uneasily upon a species ontology. As the authoritative United Nations Environment Program's *Global Biodiversity Assessment* puts it, most studies that aim to quantify and map biodiversity

> focus primarily on taxonomic group units and on species in particular, not because they are of greater significance in biodiversity terms than ecological systems or genes but because these taxonomic units can be counted and, if identified securely, summed across ecological hierarchies and across geographical scales.[25]

Species are employed as the basic units for quantifying biodiversity and constitute the principal targets for conservation action.[26] Although the UKBAP also identifies forty-five priority habitats for conservation, a subset of species is used as indicators for measuring their health and the efficacy of conservation actions.

This consensus on the use of the species is an uneasy one. There is still a great deal of discussion in systematics over the veracity of the species as the basic building block of biodiversity.[27] This discussion concerns the very existence of any definition of a species that is universal to all taxa. There is an extensive and largely unresolved literature debating the "biological-species concept," of which I provide only a brief overview.[28] Richard Primack explains how a species is generally defined in one of two ways, which he terms the "morphological" and the "biological" definitions. The morphological definition of a species, most commonly used by taxonomists, surmises that "a species can be defined as a group of individuals that are morphologically, physiologically, or biochemically distinct from other groups in some characteristic. It relies on DNA sequences to differentiate genetically between species that look almost identical." In contrast, the biological definition of species argues that "a species can be distinguished as a group of individuals that can potentially breed among themselves and do not breed with individuals of other groups." He goes on to explain that this approach is more commonly used by evolutionary biologists because "it is based on measurable genetic relationships rather than on somewhat subjective physical features."[29] In practice, however, the biological definition of a species requires information that is rarely available in the field, where species most commonly need to be identified and differentiated.

Problems arise in differentiating between both morphologically similar "sibling species" that are actually biologically different and morphologically different individuals of the same species. Matters are further confused by hybrid species—generated by mating between otherwise distinct species. This is especially common in plant species in disturbed habitats. Evolutionary biologists and taxonomists grappling with this issue have proposed many other forms of species definition, but none of these have become a universally shared mode of differentiating the basic units of biological diversity. E. O. Wilson explains that "so far as we know, no way exists to lump or to split [species] into groups except by what the human mind finds practical and aesthetically pleasing."[30] Rather than becoming paralyzed by this barrier, Malcolm Hunter suggests conservationists use the following fallback definition: "A species is what a competent taxonomist says it is."[31] In practice, therefore, the identification of a species results from the classificatory endeavors of a talented and experienced taxonomist, who might employ a range of different definitions.

Species offer conservation biologists a practical, "intuitive ontology"—to use Scott Atran's phrase.[32] Drawing on the embodied epistemology outlined in the previous chapter, we can understand how species are much easier to tune in to than dynamic, invisible, and abstract entities such as ecosystems and genes. Discrete species—especially those that can be identified by morphological/audible differences—provide accessible units for listing, counting, mapping, and auditing the success of conservation interventions. Species provide a handy index of current ecological composition for monitoring change. They are the canary in the coal mine. As I discuss in more detail in chapter 5, diffuse and nonlinear ecological processes are much more difficult to bound, define, and monitor. Common definitions of species also favor the classification of higher-order animal species with greater ecological charisma—for example, those that reproduce sexually and are more easily differentiated by the human eye. This selection of the species as the basic ontological unit for practical conservation, as in the UKBAP, can be therefore understood as the first "cut" to be performed in the biopolitics of biodiversity. In the following two chapters, I discuss some of the problems associated with framing conservation too narrowly around species composition.

DESCRIPTION

In addition to uncertainties about the universality of the concept of the species, panoptic conservation is further stymied by significant ignorance of the total number, distribution, and dynamics of species in the world. Taxonomy is the discipline responsible for naming and classifying organisms. It has a long history, a distinct geography, and a set of knowledge practices with interests and priorities that precede (and still supersede) the biodiversity crisis and its demand for global inventories. It is also a discipline in decline.[33] In 1995 the authoritative *Global Biodiversity Assessment* reported:

> The existing species record is deficient in several respects. . . . It is partial (many species have not been described), inaccurate (it contains errors of taxonomic judgment and of many other kinds), and biased (it is clearly more complete and more accurate for some groups of organisms).[34]

Respected estimates of the potential total number of species in the world range from 3 to 100 million.[35] Estimates of the number of described species also vary,[36] with a figure of 1.2 million commonly cited as the known total. One recent authoritative paper calculates a possible total of 8.7 million and suggests that some 86 percent of existing species on earth still await description.[37] As Wilson puts it, in species terms "we live on a largely unexplored planet."[38] The biopolitical aspirations of panoptic planetary management embodied within biodiversity conservation are currently thwarted by the embarrassing absence of this basic knowledge. For Wilson and others this is a travesty, and the completion of a planetary species inventory has become a rallying cry for further investment in biological and taxonomic science.[39]

There is a striking taxonomy to this partial subset of described species, with some groups being significantly better described than others (i.e., a higher proportion of their believed total have been named). In a commentary in *Nature* in 1992, the influential biologists Kevin Gaston and Robert May argue that these description patterns reflect the "taxonomy of taxonomists," which "is ill matched to the species richness of taxa and to the magnitude of the jobs remaining to be done for different groups." They note that "with respect to taxonomic attention, the average plant species does about an order of magnitude worse that the average vertebrate species and an order of magnitude better than the average invertebrate."[40] Species favored by taxonomists are generally conspicuous, abundant species with large geographical ranges and body sizes.[41] The foci and subsequent scope of taxonomy are strongly configured by the ecological charisma of the organisms to be described.

Taxonomists' preferences are also shaped by an organism's corporeal charisma in relation to the knowledge practices and affective logics of systematics (the wider field of which taxonomy forms a part). Systematists' interests extend well beyond inventory and collection and concern classification and evolution. They are interested in questions and enjoy the answers provided by groups of model organisms, as Darwin showed famously in his enthusiasm for beetles. In a letter to his neighbor, the entomologist John Lubbock, in 1854, he reveals, "I feel like an old war-horse at the sound of a trumpet when I read about the capture of rare beetles—is this not the magnanimous simile for a decayed entomologist? It really almost makes me long to begin collecting again."[42] The taxonomic archive embodies the historical legacy of these passions and interests.

SURVEILLANCE

The United Kingdom has a low density of species and a high density of interested humans. This makes it possible to estimate the total number of national species with greater confidence, although similar taxonomic partialities exist and new discoveries occur with some frequency. When the UKBAP was drafted, the total number of described species was calculated as approximately 96,000.[43] More striking partialities exist in species surveillance, the second arena identified in Figure 4, which plays a vital role in the biopolitics of UK biodiversity conservation. As I trace in more detail in the following chapter, it is through surveillance that a species can be framed as a dynamic population, evaluated and prioritized for conservation, and made the calculable subject of various forms of intervention. Surveillance provides the data through which governance is planned and audited.

To inform the UKBAP, a review was commissioned into the state of surveillance in the United Kingdom and resulting biological data holdings. The authors identify at least 2,000 different organizations, encompassing over 300,000 people, who survey with varying degrees of regularity every year. Annually, they report on the national distribution of approximately 450 species of birds, 50 of butterflies, and 750 of lichens. A further 12,000 species are surveyed more intermittently and patchily.[44] The majority of these surveyors are volunteers. The primary organizations for coordinating their enthusiasms are the British Trust for Ornithology, which is responsible for all bird surveillance and the data it produces, and the Biological Records Centre (BRC), which coordinates the work and data of eighty smaller national recording organizations that represent all groups other than birds, lichens and algae, and sea life. As well as aggregating ad hoc data, these organizations are responsible for proactive surveillance. They devise and disseminate survey methodologies and distribute technologies to enable surveyors to trap, locate, and identify a species and to tune in to its ecology. They must ensure that the entire range of a species, or at least a stratified sample of it, is covered by their surveyors. They must also be able to collect, aggregate, standardize, store, and analyze the results of any surveillance initiative. Historical enthusiasms for counting species have generated a wealth of data. The authors of the review identify 1,386 datasets, stored in over two thousand different locations and containing over 63.5 million records.[45]

Although the UK flora and fauna are arguably the most surveyed of any nation in the world, their surveillance shows a distinct taxonomy. This is also a concern for the review, whose findings are reproduced in Table 1. On a broad scale it identifies a similar taxonomy to species description, with preferences shown in the percentage of total data toward vertebrate species (66 percent) and plants (25 percent), rather than invertebrates (8 percent). The largest single dataset is the 23 million ringing records held by the British Trust for Ornithology. A closer look at the vertebrate records shows that nearly 99 percent of this data has been collected solely on bird species. The United Kingdom's 300,000 mammal records are dwarfed by the 41 million records that have been collected on birds.[46] Among invertebrate records there are distinct preferences expressed toward Lepidoptera (butterflies and moths) and Coleoptera (beetles).

Taxonomic partialities toward birds are easily linked to their ecological and corporeal charisma, outlined in the previous chapter—most especially, their relative detectability and potential for listing. The readiness of birds for surveillance also helps explain the longstanding, popular, and well-resourced institutional infrastructure that exists to support bird monitoring and research. There is a positive feedback mechanism at work here whereby potential surveyors are drawn to visible species with accessible guidebooks, sociable networks of fellow enthusiasts and mentors, and well-organized survey programs.[47] Here, past preferences come to groove future practice. Such inertia can be circumvented. For example, several of the naturalists I spoke to while conducting this research explained how the publication of an accessible guidebook to UK hover flies[48] spurred lepidopterists, dragonfly enthusiasts, and even some birders to expand the scope of their observations.

Institutional capacity for surveying a group of species also helps overcome what the sociologist of science Geoffrey Bowker has termed

TABLE 1. THE TAXONOMIC BREAKDOWN OF UK BIOLOGICAL RECORDS		
TAXONOMIC GROUP	TOTAL RECORDS ('000S)	% OF TOTAL DATA
Lower plants	2,153	3
Vascular plants	13,937	22
Invertebrates	5,314	8
Butterflies and moths	1,556	2.4
Beetles	740	1.2
Vertebrates	41,918	66
Fish	98	0.1
Herpetiles	31	0.1
Birds	41,340	65
Mammals	303	0.5
Total	63,442	100

the problem of "datadiversity" in biological records.[49] This describes the multitude of diverse data standards layered into biological databases that hinder their successful integration and deployment for conservation. This was a major problem for the UKBAP and was most acutely experienced in the data generated by the large number of relatively small datasets on invertebrates that are listed in Table 1. Different and discrete cultures of counting and collecting have given rise to multiple names and taxonomies, incommensurable time series and mapping conventions, and diverse listing practices encoded into incompatible software programs. All of these conspired to curtail the usefulness of historic records and made data management a key concern of the UKBAP, resulting in a significant investment in the creation of the National Biodiversity Network.[50]

SPECIES ACTION PLAN IMPLEMENTATION

One of the key factors that influenced which species had sufficient knowledge to be evaluated and prioritized for conservation was a species' representation by an NGO. As a result of the voluntarist, not-for-profit political economy of UK conservation in the late 1990s, the interests, efficacy, and power of conservation NGOs becomes even more important for implementing species action plans, which requires funds, popular support, political leverage, and land ownership. There is a great range and diversity here. In their historical development, UK conservation NGOs have tended to focus on either specific taxa or land acquisition. Examples of the former include the Royal Society for the Protection of Birds (RSPB), Plantlife, Butterfly Conservation, the Mammal Society, and the Herpetological Conservation Trust, to name but a few. Although these organizations do collaborate, they are often in direct competition for members and the voluntary enthusiasms of amateur surveyors. There are stark differences in the resources they can mobilize. For example, in 2001 the RSPB had over a million members and an annual income of £48 million. It also owns or manages over 100,000 hectares of land.

When it came to the UKBAP, NGOs tended to concentrate their resources on the action plans for species in their taxonomic jurisdiction. Birds were well represented by the RSPB. Their popularity coupled with the organization's sophisticated lobbying power even ensured that populations of wild birds were included in 2001 as a "headline indicator" in the (then Labour) government's "quality of life counts." These claimed to provide transparent targets to measure progress on achieving sustainable development.[51] In contrast, invertebrate groups were poorly served by the large number of small, amateur, and often impoverished organizations. There was little coordination between these groups as to how conservation status should be evaluated, priorities should be set, and who should be responsible for implementing the invertebrate action plans.[52] This state of organizational disarray helps account for the high number of invertebrate plans that did not have a lead partner organization and thus saw limited implementation. Again, past wealth and capacity unequally distributed across taxa came to groove future action, shaping the performance of biodiversity conservation.

But the past is not determinate. As a result of the organizational disarray in UK invertebrate conservation in response to the UKBAP, key figures within the statutory authorities and invertebrate world decided to form an umbrella conservation body for invertebrates. Buglife was founded in 2002 with support from the RSPB and a number of other organizations, along with two large legacies.[53] It has since flourished, building up a membership, establishing targets, bringing together and standardizing datasets, and generally raising the public profile of invertebrates in the United Kingdom. The dynamics of beetle populations are not yet taken as surrogates for the good life, but their fate is in better hands.

PERFORMING BIODIVERSITY

In this chapter I have examined biodiversity as a mode of biopolitics. In theory, biodiversity promises a new way of understanding and governing conservation. It has panoptic aspirations to secure the full diversity of life at the interconnected scales of genes, species, and ecosystems. It imagines a scientific paradise of well-funded, rational resource management, enabling the full inventory of life, systematic means to monitor its dynamics, and efficacious means to intervene at moments of crisis. In practice, as I have traced, in the context of the United Kingdom biodiversity as biopolitics works within the ecological, corporeal, cultural, and institutional constraints that characterize the assemblage of UK nature conservation. It is dependent upon the material possibilities for field science, historic data, the real estate of nature reserves, and the resources of NGOs.

Here, biodiversity as biopolitics cuts up the flux of wildlife to create a practical set of units for action. It has tended to focus on species rather than on genes or habitats. There is a clear taxonomy among these species as to the degree of interest, knowledge, resources, and institutional support they receive. Birds, large plants, mammals, and some groups of invertebrates, like butterflies and beetles, have been popular. The vast majority of invertebrates have been relatively neglected. This analysis suggests that UK conservation is guided less by the panoptic logic of biodiversity and more by a taxonomy of nonhuman charisma that emerges out of the encounters detailed in chapter 2. This taxonomy is embedded within the institutional assemblages that perform conservation. There is a grooving effect here, where past practice shapes future encounters. But there is also scope for difference. New technologies, charismatic individuals, and institutional developments have helped widen the scope of knowledge and concern. The implications of this analysis are important for understanding and informing the scope and conduct of conservation after the An-thropocene. Paradoxically, for a science and practice dedicated to a return to Nature, biodiversity is fundamentally data based—its future tied to its presence in lists, categories, and action plans. As Geoffrey Bowker notes in his work on biodiversity databases, assemblages are performative, anticipating and shaping the bodies and ecologies they purport to represent.[54] While acknowledging my caveats, biodiversity conservation in the United Kingdom is likely to result in a conver-gence between the ecologies known and valued in the current con-servation assemblage and the landscapes subject to its governance. In an epoch of accelerated extinction, the future looks bleak for the vast majority of forms of life not blessed by charisma, adaptive enough to go feral, or productive enough to be domesticated.

In the next chapter I illustrate the further significance of this per-formativity by examining a popular, successful, but contested mode of European conservation that targets populations of species that inhabit premodern agricultural landscapes. The aim of this mode of biopoli-tics is to avert change, to preserve present composition through calcu-lation and intervention. I compare this with an emerging alternative concerned less with species and past–present ecologies and more with nurturing processes. I attend to the ontological politics at the interface of these very different ways of choreographing conservation.

· 4 ·

CONSERVATION AS COMPOSITION

Securing Premodern Ecologies in the Hebrides

The corncrake is often held up as a success story of UK biodiversity conservation. In the 1980s its UK population was in seemingly terminal decline. Once common across the entire country, it had retreated to the Scottish Hebrides, where a few hundred birds spent their summer breeding season. The corncrake was one of the 391 species prioritized for action under the UKBAP. It became subject to a comprehensive national surveillance program and a detailed species action plan. Thanks largely to the efforts of the Royal Society for the Protection of Birds (RSPB), its population dynamics are now known and understood, and its decline has been reversed. The future of the corncrake and crofting (the local low-intensity agricultural system that creates its desired ecology) seem to have been secured.[1] This is a good-news story for corncrakes, crofters, and many conservationists.[2]

In this chapter I dwell on the story of corncrake conservation for two main reasons. The first is that it offers a compelling illustration of the utility of conceiving conservation as biopolitics, as I introduce in general terms in the previous chapter. Here, I detail the processes through which the corncrake and crofters were governed, working through the four arenas identified in Figure 4 (see chapter 3). I focus on the statistical framing of corncrakes as a dynamic population that could be counted and whose "productivity" could be scientifically modeled and calculated. I examine how this required the creation of field laboratories in nature reserves in the Hebrides. I then trace the processes through which corncrakes and crofters were governed to perform to the "optimized" output of this productivity model. This involved material, legal, and economic interventions into land management. Here, I expand the interest in biopolitics of the previous chapter in order to explore questions of human governmentality, as well as the politics of putting conservation science into practice.

Second, I dwell on corncrake conservation because it is exemplary of the orthodox approach to nature conservation in the United Kingdom (and much of western European) that I will term *conservation as composition*.[3] Here, science and policy are concerned with the preservation of premodern agricultural landscapes—specifically, those left behind by the long demise of feudal peasant agriculture before the radical agricultural intensification that took place after World War I. These landscapes tend to be understood as fixed and timeless places, nurtured by "traditional" farming, and threatened by both its intensification and its abandonment. Wilderness is not a priority here. It is wary of change, finds solace in the past, and is ambivalent about the future. It desires ecological (and political) order rather than surprises and is concerned with species and habitats rather than ecological processes and function.

This approach is informed by equilibrium ecology, which provides a scientific framework for the rationalization and institutionalization of conservation. Equilibrium ecology conceives of nature as a homeostatic machine. This ontology enabled the classification, control, and manipulation of an objective, balanced, and predictable nature. It informed the designation and management of nature reserves and the creation of extensive lists of priority species and habitats that could be made subject to the comprehensive monitoring and governance programs recounted in the previous chapter. These in turn became embedded within the territorial, legal, and economic assemblage of European conservation.[4] This approach has been successful, but it has its problems and its critics. I narrate corncrake conservation in this way because, in the following chapter, I compare it to an alternative model of conservation, which is illustrated (though not perfectly) in a very different example.

CORNCRAKE CONSERVATION

The corncrake has a long and multifaceted folk history. It has fascinated poets, tantalized chefs, and eluded hunters for centuries.[5] Unlike the vast majority of small, inaccessible, and indistinct species, the bird was not discovered by science. It was once a ubiquitous part of the soundscape of the UK countryside and had numerous vernacular names. The corncrake became one of the first additions to what is now the global inventory of described species when Carl Linnaeus listed the bird as *Rallus crex* in an early edition of his *Systema Naturae*.[6] It became *Crex crex* in 1803 when Johann Bechstein invented the genus *Crex* for craking birds. There are no subspecies.[7]

This long-standing interest in the corncrake among natural historians continued through the nineteenth and early twentieth centuries, when the species' decline was first noted and its extent and possible causes discussed.[8] This research was largely anecdotal, and the surveillance that accompanied it was patchy. It was only in 1985, when the corncrake was explicitly prioritized by the RSPB, that any significant efforts were made to quantify, account for, and address its decline. As I identify in the previous chapter, the RSPB is one of the United Kingdom's oldest, largest, wealthiest, and most landed conservation NGOs. Its principal and historic focus has been UK birds, though it increasingly speaks for the environment as a whole. It has focused its resources on science, land acquisition, and various forms of advocacy, generally in the interests of the wildlife that depends upon traditional agriculture and other forms of low-intensity land management. The science, politics, and economics of corncrake conservation are exemplary of its operations. In the following sections, I trace the passage of the corncrake through the four arenas for successful conservation identified in the previous chapter. This trajectory is illustrated in Figure 6.

To give an overview, surveying and researching the corncrake involved a series of closely choreographed encounters between the bird, RSPB scientists, and their technologies, including those detailed in chapter 2. These encounters can be divided into two key types that inform each other and proceed in parallel but have slightly different aims. In the first a set of nocturnal male crakes are framed as dynamic populations of the species *Crex crex* whose population multiplication rates can be calculated and compared over time and space. These calculations allow conservationists to evaluate the rarity and threatened status of the bird in relation to common criteria. In the second set of encounters in a field laboratory, the dynamics of different corncrake populations

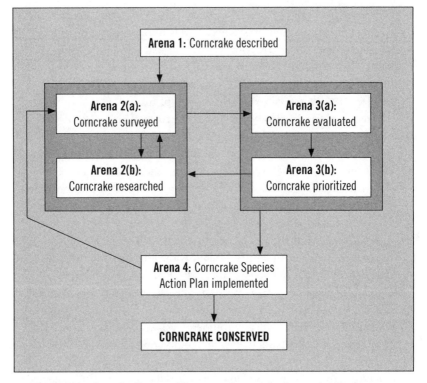

FIGURE 6. A schematic diagram of the corncrake conservation assemblage.
The arrows indicate developments over time.

are tied to certain agricultural practices, with varying degrees of confidence. These calculations generate a corncrake productivity model that can simulate likely mortality rates under different land management scenarios. The aim here is to frame the corncrake as a casualty of the intensification of agriculture. Optimizing the productivity model identifies an ideal set of corncrake-friendly land management practices to be implemented through the corncrake species action plan.

Framing the Corncrake as a Dynamic Population

Like many forms of modern biopolitics, governing the corncrake involves conducting a national census. This was first performed in 1978–79 and is now carried out quinquennially. The census is exemplary of avian surveys in the United Kingdom. It requires an understanding of the species' behavior to develop a robust census methodology, a network of skilled and disciplined surveyors covering the entire species range, and repeated counting according to the methodology. It depends on the standardized recording of these counts and the aggregation and examination of their results at a "center of calculation."[9] To prepare the census methodology, the RSPB researchers required an understanding of the behaviors and ecologies of the Hebridean corncrakes. Here, they pioneered the forms of interspecies communication detailed in chapter 2, learning to be affected by the corncrake's nocturnal call and tuning in to its diurnal, mobile radio signals. This involved the imaginative use of technology and the generation of new skills, habits, and affects and led to the creation of a new soundscape in the Hebrides. By 2000 about one hundred corncrakes had been trapped, tagged, and tracked, and records of their behavior and ecological time–space rhythms had been inscribed into numerous maps, tables, and field notebooks.[10]

Back in their labs on the mainland, the RSPB researchers set themselves two tasks. First, they gathered the quantified data from the hundred corncrakes that were tagged and tracked and built a database. Taking this hundred as a representative sample and calculating medians and averages within certain "confidence intervals," they generated a generalized understanding of the corncrakes' behaviors and rhythms.[11] This understanding informed the second task, which was to develop the census methodology.[12] This needed to provide practical, rigorous, and standardized techniques that allowed other less experienced surveyors to tune in to, differentiate, and count all of the calling male birds in their allocated survey area. They focused on male corncrakes because their calls are accessible. Females remain invisible and largely silent. The methodology is pedagogical and seeks to teach surveyors how to be affected by corncrakes, disciplining their bodies and overriding previous affections. The census methods seek to textually encode a subset of the embodied field skills that the RSPB researchers had developed during their previous summers in the Hebrides. In an ideal world future surveyors would read and learn these methods—downloading them like software—and then head out to the field as standardized, programmed scientific instruments. As seen in my confused encounters in chapter 2, governing eager but inexperienced surveyors is not so straightforward.

While carrying out their field research, the RSPB had heard only from a sample of UK corncrakes. To speak for the entire population, they needed to count the rest of them. For this they had to build a panoptic—or panauditory—surveillance assemblage, patrolled by a network of suitably qualified surveyors, adhering to the same methods. Constructing this epistemic geography was no modest undertaking. The entire known corncrake range was divided into twelve broad regions (as illustrated in Figure 7), which were further subdivided into thirty-five survey districts. This division converts the complex topography of the Highlands and the Islands into neat Euclidean subdivisions. Like the best cartography, it performs the god trick of extraterrestrial vision. This mapping is configured around practical and political geographies. Rather like postcode districts, each of the thirty-five districts is designed to be of a size that can be easily managed by one professional surveyor and take into account the existing administrative geography of the RSPB's reserves and network of wardens.

To conduct the census, the RSPB draws on its existing network of field staff, supplemented by contracted surveyors. This ensures that all but the most inaccessible parts of each survey district are visited the requisite number of times. Members of the extensive British amateur birding community are also enrolled to extend and supplement the coverage of the census. Through various media they are encouraged to call a national "corncrake hotline" if they think they might have heard or seen a corncrake. Finally, a roving band of RSPB employees travel around Scotland playing prerecorded corncrake calls over likely looking patches of vegetation. These corncrake minstrels seek to "stimulate singing"[13] among territorial males for subsequent confirmation according to the census methods.

The RSPB researchers strive to ensure the rigor of their survey by making sure that each record is compiled by a suitably qualified surveyor adhering strictly to the standard methods. This epistemological adjudication is complicated and political. It is difficult to assess a surveyor's skill at a distance; there are no standard qualifications, and abilities vary greatly between individuals. Surveying ability is judged

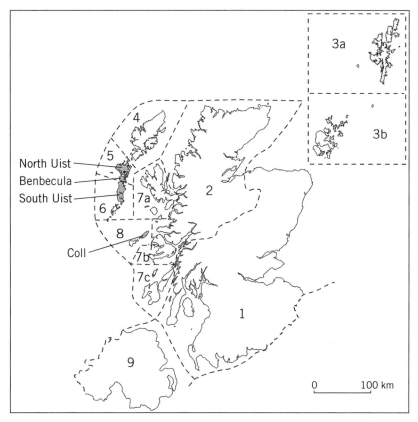

FIGURE 7. A map of Scotland showing the twelve corncrake census areas.

largely by experience and informal recommendations. For example, to verify the records received on the corncrake hotline, promising callers are connected to their local RSPB office, where an employee quizzes them on what they have heard and tests their general ornithological knowledge. UK birders are a proud and enthusiastic flock and do not take kindly to being excluded or, even worse, told that what they thought they heard was not a corncrake. For a membership organization such as the RSPB, it is important to let down enthusiasts gently. Furthermore, professional surveyors have their own idiosyncratic ways of birding. Making surveyors adhere to the methods is as much about ironing out previous modes of engaging with birds as it is about forging a new way of being affected.

Having established the census methods, divided the field, and distributed it among skilled and disciplined surveyors, the corncrake census can now take place. This involves repeated acts of nocturnal corncrake counting like those detailed in chapter 3. Surveying corncrakes, as Craig happily admitted, involves informed guesswork and the ingenious bending or customization of the rules laid down in the standardized methodology. For instance, in order to cover the large area of land he has been allocated in the time available, Craig has developed a method of listening while driving. The census says he should stop and walk, but driving with all of the windows open, he has learned to identify corncrakes above the noise of the car, the passing wind, and the surprisingly similar noise made by his car cruising past a standard Hebridean fence. Like the field guides described by Michael Lynch and John Law or the schema for classifying vegetation employed by Clare Waterton,[14] Craig uses the census methods as a loose script or "sensitizing device" to guide his field performance.[15]

Here, Craig is practicing an archetypal form of field science.[16] He must accommodate himself within and tailor his work to the ecological and social dynamics and idiosyncrasies of the Hebrides. In contrast to a generic, orderly laboratory, where the inscribed corncrakes will shortly end up, in the field Craig must negotiate with the crofters whose land he is surveying. They were initially (and perhaps understandably) alarmed by his nocturnal activities and were none too receptive to the meddling of "incomers." Craig has socialized the scientific methodology, spending time in the day forewarning those he would encounter later and avoiding land owned by the most obstreperous crofters. He must also deal with the vicissitudes of the weather. The Islands are normally windy and often wet; when I visited in 2003, the wind rarely seemed to drop below a force 5, which is the official cutoff for surveying. It is rare that the same weather conditions last for a whole day, let alone for a duration that would allow their effect on nocturnal listening to be externalized from this field experiment. Craig must calibrate his hearing to the wind, embodying and localizing the survey methodology through distinct "practices of place."[17]

Once he has heard a call and tracked it to its location, Craig plots the individual bird by flashlight onto a photocopied map of the island. This male corncrake is immediately given a discrete and spatial identity: a code that ensures its "reversibility" in its relation to its referent bird.[18] Once back in the comfort of his own home, Craig transfers the locations of his birds to a master copy of the map. At the end of the recording season, he enters into a simple spreadsheet the time, frequency, and location of each bird call. Until recently, this generated a classic paper table. Now, it takes a digital form. If his Internet connection is working, Craig emails the data to the census coordinator. Often, this is not possible, and the inscribed corncrakes depart by mail on an airplane.

As their fleshy parents leave for Africa at the end of their Hebridean breeding season so their encoded carbon or silicon offspring depart for the mainland. They wing their way to Rhys Green—the principal research biologist for the RSPB and corncrake census coordinator. In his laboratory in the zoology department at Cambridge University, he collects the spreadsheets on his desk, enters them into a computer, queries any peculiarities or absences, deciphers handwriting, and fills in the gaps. The lab is a much more amenable place to understand corncrakes—it is private and climate controlled, and the birds are all present and visible and can be combined together. There are no crofters preventing access or asking difficult questions. As Wolff-Michael Roth and George Bowen put it, the corncrakes have been "domesticated."[19] It does not matter that the real birds are miles away in southern Africa. Rhys Green can relax and do science.

The science that he wishes to do at this stage is reasonably straightforward. Aggregating all the counted corncrakes, he creates a database. This represents the completion of the panoptic gaze and, for the first time, brings all the male corncrakes in Britain together in one place, framed as a population. Holding the population within his database, he first sums the individual inscribed encounters to generate an annual total for the entire population, broken down by different areas. Comparing the latest population totals with those from previous censuses, he performs a neat statistical transformation to give an indication of the population dynamics of the species, both over its whole range and in different areas. These translations calculate the mean population multiplication rate (PMR) between two survey years; they frame the corncrake as a *dynamic population*.[20] In short, the figures show that the UK corncrake population was declining until 1993 but has since rallied and is now on the increase. The PMR has become the benchmark figure against which the RSPB can monitor the fate of the species, identify causes of decline, and audit the efficacy of future conservation interventions.

Framing the Corncrake as a Casualty of Changing Agricultural Practices

The link between agricultural intensification and the decline in the British corncrake population had been suspected since the 1940s.[21] There was a clear and widely accepted hypothesis but no scientific proof. Without proof it was difficult to leverage political action. Rhys Green and his fellow RSPB researchers therefore needed to conduct an experiment. Here, an experiment constitutes a scientific procedure undertaken to test a hypothesis or demonstrate a known fact. It is performed with clear intentions and an expected set of outcomes and is framed within a general theory. Sympathetic commentators in the philosophy of science present the ideal experiment as structured by the hypothetico-deductive method. As Robert Kohler notes in his writings on the lab–field border, such controlled "experiments in nature" (i.e., in the field) are rare. Most field science involves what Kohler terms "nature's experiments"—where scientists observe, measure, and model natural variation, accommodating themselves to the field without interfering with (and thus contaminating) its objective (found) reality.[22] To date, the RSPB scientists had been engaged with nature's experiments; to experiment in nature they needed a laboratory.

Modern science holds the laboratory as the optimum "truth spot" or "locatory" for knowledge production.[23] Laboratories are private and artificial (made). As with Rhys Green's office at Cambridge, they allow scientists to domesticate the wild, securing total control over the presence and behavior of their research participants—both human and nonhuman. Laboratory walls render science inconsequential in the sense that they establish a clear spatial division between a knowledge object and the world it purports to represent.[24] In contrast to the specificities of the field, the strict standardization of laboratory conditions allows scientists to generalize that knowledge generated "here" applies "everywhere."[25] In identifying their laboratory for their experiments in nature, the RSPB scientists focused their attention on the islands of Coll and North Uist (see Figure 7), where the organization owns nature reserves. These nature reserves are liminal spaces in relation to the distinctive epistemic properties of the lab and field. They are privately owned and are farmed on contract but are accessible to RSPB members and the paying public. They are authentic, found—i.e., natural—sites but can be subject to forms of manipulation in the interests of research and practical conservation. Nature reserves seek to publicly demonstrate good management, and thus these interventions seek to have consequences. They are specific, exemplary sites of machair habitat, but as laboratories, their findings must be made to stand in for corncrakes and crofters everywhere.

To conduct their experiment, the RSPB researchers caught, tagged, and tracked the activities of the corncrakes resident on their reserve laboratories.[26] Listening and observing, they noted the timing and duration of their breeding and chick-care cycles, their egg clutch sizes, and their background mortality rates. In parallel they also tracked the corncrakes' interactions with the local crofters and their agricultural machinery to explore the mortality rates of clutches of eggs and broods of chicks under different mowing regimes. Crofters were instructed, paid, and supervised to cut as usual,[27] but at different times and with different mowing patterns to simulate the effects of the intensification of agriculture—specifically, the advent of more powerful mechanical tractors and the shift from making hay to earlier and more frequent cutting for silage. In the lab crofters could be disciplined, incentivized to put aside the economic logics of efficiency and profit and rewind the history of recent agricultural development. The scientists identified that the later in the year a field was cut, the greater the number of corncrakes escaped the mower. They also established that corncrakes tended to flee the mower in a perpendicular direction and became trapped in the ever-decreasing circle when the field was cut from the outside in.

To assess the statistical significance of these findings and, thus, to test and hopefully prove the hypothesis, the scientists built a mathematical model. Here, the inscriptions of the reserve corncrakes were combined with further data on corncrake vegetation preferences and crofters' mowing dates and cutting regimes generated by the corncrake surveyors. These data were statistically translated into a set of probabilities, averages, and constants, which were used as a model of "corncrake productivity."[28] This model is a quantified measure of the number of chicks reared to independence per female in a breeding season. It takes the form of a flow diagram with various feedback loops. Values for the key variables are entered, and the model is run recursively to encompass an entire breeding season. As it runs it keeps a dynamic total of corncrake productivity. By changing the values of the key variables associated with agricultural intensification, it was possible to prove a net decline in corncrake productivity within certain confidence intervals. The model and the evidence on which it was based were published in a prestigious journal and confirmed the long-suspected hypothesis that the corncrake was a casualty of changing agricultural practices.

The second key role of the model was to be performative. It allowed the corncrake researchers to abstract themselves from the messiness of the field and its constitutive negotiations to design and anticipate the impact of future scenarios. Its inventors note that the model was constructed specifically to investigate agricultural practices that were "susceptible to manipulation by conservation bodies"—namely, cutting less, cutting later, and cutting in a different fashion (from the inside out).[29] With these in mind, the researchers optimized the model to identify the ideal combination of values for these variables that would maximize corncrake productivity. This combination helped identify a set of corncrake-friendly land management practices[30]—including a distinct mowing regime (Figure 8)—whose implementation became one of the main aims of the corncrake species action plan.[31] This was drawn up by the RSPB, who became lead partner for the corncrake in 1995.

Implementing the Corncrake Species Action Plan

On the strength of these calculations and the quality and volume of data by which they were informed, the conservation status of the corncrake could be easily evaluated. It was one of a small number of

Mowing towards rocky knoll to leave a sizeable area of unmown grass.

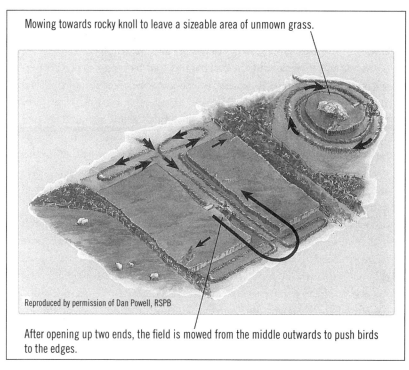

Reproduced by permission of Dan Powell, RSPB

After opening up two ends, the field is mowed from the middle outwards to push birds to the edges.

FIGURE 8. A diagrammatic illustration of the corncrake-friendly mowing regime developed out of the RSPB research. Image reproduced courtesy of Dan Powell, RSPB, www.powellwildlifeart.com.

the 391 UKBAP priority species with sufficient data to allow a national evaluation. The corncrake was declared "nationally scarce."[32] The same data were also passed to Birdlife International, who was the IUCN Red List Authority responsible for evaluating the global conservation status of bird species. They drew on their international database to evaluate the corncrake as "vulnerable" and therefore "globally threatened."[33] These rankings ensured that the corncrake was listed in the schedules and annexes of the relevant UK and European conservation legislation—most important, the EU Birds Directive. This legislation established governments' legal obligation to protect present corncrakes and secure their future.

To implement the corncrake management plan, RSPB employees

needed to scale up and roll out the corncrake-friendly land management practices that their scientists had developed on their reserve
laboratories over (and ideally beyond) the entire corncrake range.
Corncrakes, crofters, and their ecologies had to be governed to perform to the optimal model scenario. This involved a series of political
and economic negotiations. As with many of the UKBAP priority
species, the corncrake has a distinctly hybrid ecology that is shaped
by and dependent on forms of premodern land management. European corncrake populations grew significantly in the five thousand
years prior to the Industrial Revolution. During this period extensive
anthropogenic deforestation and subsequent burning created vast ex-
panses of suitable grassland habitat, which persisted with the development of low-intensity arable systems across Europe. RSPB research
had demonstrated that the powerful mowers and earlier and more
frequent cutting associated with agricultural intensification makes
this habitat uninhabitable. Although research on the ecological consequences of the demise of collectivized agriculture in eastern Europe
suggests that the corncrake benefits temporarily from land abandonment, the habitat soon becomes unsuitable as dense scrub develops.[34]
Like the Asian elephants in Sri Lanka encountered in chapter 1, the
corncrake has flourished as a consequence of past human impact.

   To maintain corncrakes at this (perhaps historically anomalous)
population high point, the RSPB had to preserve a particular political
ecological status quo. The small pockets of suitable corncrake habitat in the United Kingdom persist because the low-intensity land-use
practices in these areas still approximate those of the corncrake's premodern boom time. As with many of the RSPB target bird species and
the wider priorities of UK conservation, the future of the corncrake
has thus become tied to the survival or simulation of economically
marginal forms of agriculture. As such, RSPB employees have had to
work closely with the Hebridean crofters and farmers who manage the
vast majority of the UK corncrake habitat and a wider set of actors
who control the political economy of European agriculture.

   As sociologists of the field sciences have frequently noted, deal-
ing with farmers poses a series of epistemic, cultural, and political
problems to environmental scientists.[35] When venturing out of their
urban offices and laboratories, the RSPB wardens and outreach offi-

cers quickly learned they could not just turn up in white coats with published science and expect to direct proceedings. Initially, the RSPB delegates appealed to the crofters on moral terms. In person and through leaflets and workshops, they tried to persuade them that they had a cosmopolitan obligation toward the threatened corncrake on behalf of the nation and even perhaps the world.[36] The crofters contested this mode of environmentality on epistemic grounds. They argued that the corncrakes were there only because of their natural knowledge and care, so why should they change their practices and incur additional expense? This epistemic politics was given further freight by challenging the implication that they should take the advice of a powerful outsider from "down south." Early RSPB wardens were most commonly from England and apparently rather naïve to the fraught political history of the Highlands and the Islands. The tragic story of the Highland clearances and the violence inflicted on crofters by powerful absentee elites is long, bitter, and too extensive to detail here.[37] In its increasingly prevalent nationalist rekindling, it provides a series of powerful discourses to oppose "incoming" power and expertise.

Crofting—like much farming in upland Britain—is, however, an economically marginal practice. It is maintained largely by subsidies from the EU Common Agricultural Policy (CAP) as directed and delivered at a UK scale. This marginal and dependent position presented the RSPB with other opportunities for intervention. In 1992 the RSPB initiated a grant scheme called the Corncrake Initiative. A new generation of wardens, more attuned to local sensitivities, approached crofters with a confirmed corncrake on their land and offered to meet the additional costs of complying with the corncrake-friendly land management practices. A further scheme offered funding for the proactive creations of "corncrake corners" or "corncrake corridors" in areas in which research suggested might be suitable. These schemes were popular, but their success increased the financial burden on the RSPB, which began to look to leverage other public sources of funding.

During the late 1990s and early 2000s, a series of EU and UK taxpayer-funded agro-environmental subsidy schemes began to come on stream in Scotland. These emerged from reforms to the CAP in response to criticisms by environmental NGOs (including the RSPB) and others about the environmental damage wrought by the priority given to maximizing production. The reforms sought to "decouple" subsidy from production and offer support to environmentally friendly farming practices. The most important in the Hebrides was the Environmentally Sensitive Areas (ESA) scheme, which targeted areas of environmental significance, including the habitat of 84 percent of the UK corncrake population.[38] The ESA is a voluntary scheme. Payments for favorable management come in four tiers, where first-tier payments are for adhering to basic management prescriptions and subsequent tiers involve more complex prescriptions and more generous funding. As a result of extensive lobbying by the RSPB in Edinburgh, many of the corncrake-friendly land management practices that had been developed and tested under the Corncrake Initiative were incorporated unchanged into the eligible body of ESA land management prescriptions across a wide area. This shifted a significant proportion of the financial burden of implementing the corncrake action plan onto the taxpayer. It also ensured that corncrake conservation became configured around the anticipated "likely presence" of a bird. As in Steve Hinchliffe's analysis of the political ecology of UK black redstart conservation, funding became available to support corncrake conservation without the need for a survey to formally make the bird present.[39]

RSPB Scotland employed a full-time agricultural policy officer. It was her job to lobby the minister responsible for the Department for Environment and Rural Affairs at the Scottish Executive for conditions favorable to the implementation of the corncrake-friendly land management practices. She made good use of a range of formal consultations and informal networks, including frequent and cordial collaborations with the Scottish Crofters Federation. In her advocacy she placed a great deal of emphasis on the importance of the scientific research published in peer-reviewed journals, whose generation I have discussed. Campaigning materials explicitly mobilized the PMR data to graphically contrast the differing fates of corncrakes in areas with and without an ESA. The normative tone of the advocacy document was supported by many of the journal articles themselves, which in various fashions argued for public funding for corncrake conservation. This lobbying was backed up by the threat of infraction—reminding policy makers of the legal status afforded the corncrake by virtue of its listing on the EU Birds Directive.

While these subsidy schemes were being developed and implemented, the RSPB was actively acquiring and ensuring the legal designation of land for corncrakes. The listing of the corncrake on the flagship EU Birds Directive created a legal responsibility for national statutory authorities to designate any site containing more than 1 percent of the UK corncrake population as a Special Protection Area and to subsidize and oversee their sympathetic management. Ten SPAs were designated for corncrakes, encompassing about 42 percent of the total UK corncrake population. A more coercive strategy employed by the RSPB involves buying the land inhabited by corncrakes. Six of the RSPB's 150 reserves in the UK currently accommodate some 10 percent of the total UK corncrake population.[40] In 1991 they bought a 1,221-hectare farm on the island of Coll explicitly for the corncrake.

Performance

The result has been a remarkable and rare conservation success story. The decline of the corncrake has been halted and reversed, and the RSPB are now collaborating on a proactive reintroduction program on one of their reserves in Cambridgeshire.[41] The RSPB have successfully mobilized the corncrake as a flagship species, uniting the disparate epistemic communities of crofting and conservation to secure the precarious economic future for the political ecology of the Hebrides. This is a remarkable scientific and biopolitical achievement. For the first time UK corncrakes have been aggregated as a quantifiable, dynamic, and knowable population of a species that can be made subject to rational management. They have proved beyond a reasonable doubt that the corncrake is a victim of agriculture by bringing the field in all of its specificity to the lab and then returning the lab to the field in the space of the nature reserve. They have generated knowledge about corncrakes, crofters, and their ecology that applies everywhere. Informed by this knowledge, the organization has successfully navigated their flagship through the fraught, murky, and byzantine bureaucracy of European agricultural and environmental policy to keep crofters crofting. Order is maintained; the power of science is displayed; and an equilibrium ecology is preserved.

In a celebratory (and perhaps rather self-indulgent) article, Rhys Green reflects on the extent and efficacy of this achievement by testing the performance of the model against later census data on the behavior of corncrakes and crofters. He explains that the various subsidy schemes have led to real changes in crofters' practices. The average mowing date in the corncrake areas has been moved back by nearly two weeks, and most crofters have switched to the inside-out corncrake-friendly mowing regimes. Based upon these changes, the model would predict a PMR increase from 0.965 to 1.141. The corncrake census results show that the PMR of counted corncrakes is close to this estimate at 1.09. This is close enough to allow Green to conclude that the "recent reversal of the long-term decline of the Corncrake in Britain is in accord with [the model's] expectation."[42] Full marks for performativity.

SOME PROBLEMS WITH COMPOSITION

Corncrake conservation was so successful because it managed to foreclose on other possible futures for the Hebrides and ways of knowing about them. It presented and delivered a compelling and popular vision for preserving the political, economic, ecological, and scientific status quo. It has hitherto prevented both the intensification of agriculture and its abandonment. Both of these scenarios would have generated very different ecologies less hospitable to corncrakes. The ecologies associated with the former are familiar across the rest of the country, and those of the latter, less so. There is, however, a growing interest in European nature conservation in a planned deintensification, or rewilding, of land management.[43] I outline this model in the following chapter. Here, it is important to understand how it in part emerges from a series of critiques of conservation as composition that is exemplified in the case of the corncrake.

Carol Morris and Matt Reed suggest that agro-environmental schemes are complicit in the "McDonaldization" of the countryside through the rationalization of farmland nature conservation.[44] They argue that these schemes perform all four dimensions of George

Ritzer's McDonaldization thesis.[45] In other words, agro-environmental schemes and similar conservation subsidies rationalize nature by providing *efficient, calculable* means to *predict* and *control* farmers and wildlife in order to maximize the production of biodiversity— understood as the populations of target species like corncrakes. This logic is akin to maximizing the production of cattle to make cheaper burgers. Indeed, some conservationists even spoke (only half in jest) about designing a scheme that offered crofters a headage payment per corncrake produced. This rational bureaucratic model has many advantages and successes, but it also has its dark side. In Max Weber's terms, it risks "disenchanting" and deskilling farmers through enforcing a strict adherence to the instrumental logics of ecological science. Furthermore, the blanket prescriptions of these schemes often efface and thus fail to deal with the inherent variability and specificities of the nonhuman world and local environmental knowledge.[46]

Critics argue that this model takes as its baseline an arbitrary and impoverished set of ecologies already significantly degraded by human activity. Premodern agriculture did not enhance European wildlife; the wildlife we value survived in spite of agriculture.[47] Conservation has put the cart before the horse in a fashion that would be taboo for campaigns in the tropics, for example. Seeking to preserve the agricultural wildlife of Europe would be like conserving the agricultural systems that came to replace a tropical rainforest and holding back the forest's regeneration. By pinning conservation legally to the delivery of species' populations and static habitats, landscapes get frozen in a past, and the ecological and evolutionary processes that species and habitats embody get neglected. Such biopolitics is anachronistic and risky for governing ecologies that must adapt to accelerating climate change.[48]

Epistemologically, this model is premised on equilibrium ecology. It is fixed on a set of "climax communities" that are understood to result from the linear processes of secondary succession following human disturbance. These climax communities provide the stable archetypes toward which management can be directed and against which it can be audited. Such archetypes provide fixed points to guide field deductive experiments. When conservation science aims to confirm an equilibrium, it is unable to anticipate, detect, and learn from the inevitable surprises that come from ecological adaptation.

In framing knowledge generation around the testing of preformed hypothesis, it is poorly equipped to learn from unlikely events. This framework is out of alignment with the understandings of nonequilibrium ecology that are increasingly ascendant in the science of conservation biology.[49]

Economically, this agricultural model for conservation is extremely expensive. It is reliant upon subsidizing forms of production that are not economically viable in a competitive global economy. Even with the subsidies, it is currently cheaper to produce lamb for the Scottish market in New Zealand than it is in the Hebrides. This is often wasteful as well as being ecologically damaging, as—in spite of decoupling—the subsidy system still rewards production. One primary rationale for CAP has been to keep people on the land and avert rural depopulation. In the Hebrides this has been politically successful, but many of the most valued habitats in the United Kingdom are the result of historical forms of management associated with violent and unjust acts of dispossession and deeply unequal structures of land ownership and accumulation. At times the CAP perpetuates these structures. In general it offers limited means for addressing their legacies of grievance and disadvantage. Finally and perhaps most significantly, even with success stories like the corncrake, this is a mournful, nostalgic model for conservation. There is no vision for the future here, just the reactive management of extinction. For its critics this mode of conservation seeks to "render the present eternal."[50] It lacks the ambition, hope, and vision necessary for conservation in the Anthropocene. Some believe they have found another way.

· 5 ·

WILD EXPERIMENTS

Rewilding Future Ecologies at the Oostvaardersplassen

In many ways the Oostvaardersplassen (OVP) is an unlikely icon of European wildness. This state-owned polder just north of Amsterdam was reclaimed from the sea in the 1960s and earmarked for industrial development. It is part of the largest artificial island in the world, kept afloat by dykes and continuous pumping. It is located in the Dutch suburbs, bisected by road, rail, and other infrastructure, and surrounded by some of the most valuable agricultural land in Europe (Figure 9). For a range of economic and hydraulic reasons, building never began. Instead, the site was abandoned and was colonized by wildlife, including a large flock of migratory geese. Their grazing habits kept the grassland open and helped generate a novel ecology rich in bird life. In the 1980s the management of OVP was taken over by Staatsbosbeheer (SBB), the Dutch statutory conservation authority under the direction of Frans Vera—a charismatic and controversial paleoecologist and conservationist.

From his observations of the geese at OVP, Vera began to develop an alternative understanding of the paleoecology of Europe during the Holocene that would inform his approach to conservation.[1] He suggested that the climax ecological community was not an equilibrium closed-canopy forest but a shifting mosaic of forest–pasture landscapes kept open by herbivores. He argued that European conservation should not support premodern agriculture to deliver valued biodiversity—like the crofting and corncrakes. Instead, marginal land should be taken out of production and restored for the future. The accidental ecology of OVP offered Vera the chance to experiment

FIGURE 9. An aerial photo of the Oostvaardersplassen (OVP) showing its suburban location. Source Google Earth. Image copyright 2014 Aerodata International Surveys.

and test his hypothesis. Inspired by the geese, he introduced deer and hardy breeds of cattle and horses to the fifty-six-square-kilometer reserve (Figure 10). As grazing tools, these animals were encouraged to "dedomesticate" themselves from modes of agricultural management and discover behaviors and social dynamics that would catalyze ecological processes. He hoped that their grazing and carcasses would create a novel ecosystem, full of surprising and unprecedented events. There was to be minimal human management of this "Serengeti behind the dykes"—none of the action plans, targets, or other bureaucratic technologies encountered in chapter 3.[2]

News of OVP has traveled far and wide. It has become a site of pilgrimage for European and North American conservationists and has inspired a range of experiments in other locations.[3] These seek to open up forests and alter the grazing of formerly agricultural landscapes with various combinations of large herbivores. For advocates OVP addresses many of the problems of conservation as composition that I summarize at the end of the previous chapter. It focuses on

FIGURE 10. Large herbivores at the OVP. Photograph by GerardM, Wikimedia Commons.

processes and allows scientists to explore nonequilibrium ecological dynamics in the past and future. It enables ecological restoration that helps to understand and plan adaptation to climate change.[4] It also delivers a wider range of hydrological and climatological "ecosystem services"—like storing carbon and preventing flooding.[5] For conservation organizations like the fast-growing NGO Rewilding Europe, it is exciting, ambitious, and future oriented, promising "wilder" landscapes (and sometimes "wildernesses") full of charismatic animals. For administrators in a political-economic climate of austerity, it is cheap, requiring fewer subsidies.[6] Vera and his collaborators' thinking and advocacy on behalf of the OVP experiment helped drive a paradigm shift in Dutch conservation toward "nature development," engineering "new nature" with large herbivores in a networked "ecological main structure" (*ecologische hoofdstructuur*).[7]

Vera's theory and the management of OVP have proved controversial. Paleoecologists dispute his interpretation of the archeological evidence and defend the "high-forest" model.[8] Compositional conservationists worry about the absence of a management plan with specific targets to guide progress toward a preordained ecological archetype. More specifically, they are concerned that the pressures imposed by the large herbivores will threaten local populations of rare species (especially birds) whose future they fought to secure through legal designation and agricultural subsidy. They dispute the authenticity of restored ecologies. Farmers in marginal areas are concerned that a shift in the subsidy environment away from agriculture will force them to abandon farming and their land. Fears have been expressed that this model could offer a convenient gloss for cutting expensive subsidies, waiving restrictive conservation legislation, and allowing the accelerated implementation of markets in ecosystem services. Meanwhile, animal welfarists contest the wild status of the formerly domestic cattle and horses, which go hungry and occasionally die during harsh winters. They point out that this is not the Serengeti and argue that these animals remain the property of the conservationists, who have a responsibility toward their care when undertaking any experiment. The management of the OVP has been subject to two inquiries by international commissions assembled by the Dutch government.[9]

WILD EXPERIMENTS

In this chapter, I examine OVP as a popular and influential example of the recent and wider enthusiasm for rewilding that is taking place across international nature conservation.[10] This is a new paradigm broadly oriented toward restoring species and, more important, ecological processes that are absent from contemporary landscapes—especially predation, grazing, succession, dispersion, and decomposition. Rewilding comes in many forms, with important differences between and within North American and European imaginations. Together, they share a desire to shift the target baseline for conservation away from premodern agricultural archetypes toward the prehistorical ecological conditions that characterized the northern hemisphere at the end of the Pleistocene (circa 10000 BP). The aim is to create analogs of what emerged after the retreat of the glaciers and before agriculture, forestry, and animal domestication. It promotes the "sparing" of (largely marginal) land rather than the "land sharing" of the composition model.[11] It implies an unspoken commitment to agricultural intensification and/or global outsourcing.

Rewilding offers a clear alternative to the type of conservation outlined in the previous chapter. Here, I focus on some of the controversies associated with the OVP example as a way into the wider debates about its veracity and merits and to illustrate its differences from conservation as composition. In so doing, I outline the parameters of a different mode of biopolitics for conservation, which I term *wild experiments*. I present the concept of a wild experiment in the introduction. Here, I focus on the term *experiment* because it is the claimed status of OVP as an experiment that is at the heart of many of its most controversial dimensions. I should be clear from the outset that I don't think OVP (or rewilding) is the panacea, nor is it exemplary of the desired wild experiment outlined earlier. I take the controversies around OVP as generative events, both conceptually and politically. Through disentangling their dynamics, I hope to offer some useful concepts for conservation after the Anthropocene.

In the previous chapter I introduce the understanding of an experiment most commonly associated with modern natural science and illustrate how it informed the conduct of corncrake science. Here, an experiment is undertaken to test a hypothesis or prove a known fact. It is performed with clear intentions and an expected set of outcomes and is framed within a general theory. This type of experiment is conducted rarely in the field and is partially possible only in strictly controlled nature reserves, like those described in the previous chapter. It is best done in a laboratory, where scientists can domesticate the wild and control public access to create generic, standardized spaces. The walls of the lab prevent experiments from having consequences on the wider world they purport to model, while the strict standardization of laboratory conditions allows scientists to generalize that knowledge generated "here" applies "everywhere." Through this process, Science provides objective facts to inform policy.[12]

This understanding of an experiment has been subject to a range of criticisms within the sociology and philosophy of science. Critics have argued that the ubiquity of modern science—in terms of the knowledge it has created and the unruly forces it has unleashed—means that the laboratory has taken over the world.[13] In the context of global risks like climate change and the influence of science on public policy, they argue that the Anthropocene is a global experiment, happening in real time at a scale of one to one.[14] All of us are now involved in the experiment and should be (but are often not) involved in deliberating as to its conduct and consequences. They argue that this shift necessitates a different mode of experiment, closer to that which is more often associated with the field sciences. Here, control and exclusive access are rarely achieved; various publics must be involved; and scientists must work carefully with the immanent properties of the ecology under investigation, often without any clear hypothesis or theory to test.[15] Knowledge is inductive and place specific. Authority is tied to being here.

Here, an experiment is better understood as a course of action tentatively adopted without being sure of the eventual outcome. It is an uncertain and open-ended set of practices likely to generate surprising results. One of the best-known articulations of this epistemology can be found in the work of Hans-Jorg Rheinberger, who defines an experiment as a trial or a venture into the unknown.[16] Rheinberger presents science as speculative and argues that a well-designed "experimental system" is capable of generating and detecting difference, not confirming what is known. This approach has informed work in political science advocating for the "public proceduralization of contingency"[17] in the knowledge practices of ecological restoration. Geographers like Steve Hinchliffe and Sarah Whatmore have extended this experimental ethos to human–nonhuman interactions and an appeal for a "careful political ecology": a mode of field science that remains open to the "likely presences" of nonhuman "wild things" within experimental "spaces for nature."[18] Rather than seeking to test explicit theories and hypotheses framed by transcendent archetypes of Nature, the research behind these wild experiments is interested in emergent properties. It involves an experimental epistemology grounded in an acknowledgment of uncertainty and contingency.

Michel Callon and his fellow researchers have explored the various techniques through which publics can be involved in such multinatural experiments, which comprise multiple forms of expertise, with no clear division between lab and field and where multiple futures are possible.[19] In a critical and normative intervention, they differentiate "research in the wild" from "secluded research." The latter, they argue, is most commonly associated with the lab (though it can take place in the field) and has tended to cut itself off from the publics it subsequently affects. Such secluded research still has an important role, but they argue it should be linked to its publics through engaging in research in the wild. The character and whereabouts of the wild remain rather undefined in their account, though this is not an antimodern appeal for wilderness. They are clearer as to the nature of the research that should be done in the wild and the relationships that should be established there between science and politics. They promote techniques for "dialogic democracy" that can "facilitate and organize an intense, open, high-quality public debate"[20] among emergent collectives of experts.[21]

Callon et al.'s concept of the politics of research in the wild is concerned largely with deliberations between people and technologies. It grants limited agency or political status to wildlife. As I illustrate in the analysis that follows, we can supplement their useful concept of research in the wild through an engagement with recent work offering normative more-than-human approaches to the biopolitics of conservation outlined in the introduction. In different ways these take issue with the closure inherent to the compositional model and offer ways of securing life understood as a dynamic set of processes.

This brief exegesis helps differentiate two common understandings of experiment and experimentation, which are summarized in Figure 11. The two columns in this table sketch ideal types. No pure manifestation exists, and we should not necessarily understand wild experiments as the superior alternative. Instead, the variables listed in the rows of this table detail different characteristics that are conjoined in any real-world experiment and according to which we might start to classify and critically analyze their conduct. In the rest of this chapter, I wish to illustrate the utility of this framework for interrogating the biopolitics of conservation by identifying three axes for inquiry. The first axis, entitled *found–made,* thinks through and beyond the ontological commitment to Nature that configures the contrasting epistemic properties of the laboratory and the field. The second, entitled *order–surprise,* examines the epistemological and political challenges associated with anticipating surprises and generating emergent knowledge. The third axis is entitled *secluded–wild* and develops Callon et al.'s framework for engaging publics in political decision making. In the rest of this chapter, I deploy these three axes to examine the type of experiment that is being conducted at the OVP, paying particular attention to how this differs from the mode of conservation outlined in the previous chapter.

Found–Made

The first theme develops and tries to think beyond the distinction between the laboratory as a "made" space for controlled experiments whose findings are universal and field sites as specific, authentic places "found" by scientists. In conducting and describing the rewilding experiments at OVP, Vera and his colleagues shuttle between and have more recently sought to go beyond these two positions. On the one hand, they present OVP as an ideal laboratory to test a scientific hypothesis. The land was literally made, created from the sea as part of the largest artificial island in the world. Bereft of any material cultural history, the terrain and hydrology can be sculpted with dikes, pumps, and diggers. As the site is fenced and entrenched, flora, fauna, and human access can be controlled, as with the RSPB's reserves in the Hebrides. Within these bounds prehistory can be simulated.

On the other hand, the scientific legitimacy of OVP as a site to test Vera's paleoecological hypothesis (and from which to scale up its outcomes) requires that it be accepted as analogous to wild "found" sites (past and present). Here, it is necessary to downplay human intervention and to stress the abandonment of the land, the "self-willed" or "spontaneous" nature of its ecology, and its subsequent discovery by conservationists. Histories of the site therefore ascribe great agency to the geese (and subsequent herbivores) as architects of ecological change and downplay the role of human infrastructure.[22]

Critics of the OVP experiment have tended to focus on revealing paradoxes that undermine its natural/found or social/made status. For example, commentators sympathetic to farming dwell on fences and flood control, arguing that the artificiality of OVP undermines its authenticity. The presence of such technology places rewilding on a

	EXPERIMENT	WILD EXPERIMENTS
Ontology	Transcendent order of Nature and Society	Immanent and indeterminate world of humans and nonhumans
Epistemology	Hypothetico-deductive method	Designed to generate surprises
Politics	Delegative: science creates facts, politics decides what matters	Dialogical: emergent collectives for generating and deliberating knowledge
Location	Laboratory (and occasionally the field)	The "wild"

FIGURE 11. Key properties of two models of an experiment.

continuum of modes of management, debunking claims to qualitative difference from the compositional model and thus from low-intensity agriculture. In contrast, Dutch and UK ecologists take issue with the presentation of OVP as a lab. They challenge the degree of control that has been exerted and the extent to which its findings can be generalized. They note the arrival of pollen and invertebrates from outside the reserve, flag the unusually high fertility of the soil, and highlight the unique circumstances that allowed the site to evolve.[23] OVP is presented as a distinct place, not a generic laboratory.

Partly in response to these criticisms (and with a certain degree of reluctance on the part of Vera), advocates have sought to move beyond a lab–field binary. They pitch OVP and rewilding in general as a model for conservation in the Anthropocene, where found–made distinctions hold less sway. Vera no longer presents his paleoecological baseline as an authentic return but as a dynamic "reference" for future management.[24] Emma Marris heralds the OVP experiment as exemplary for conservation in the novel ecosystems of an impure "ragamuffin earth."[25] She celebrates its feral, suburban location and supporting infrastructure. For Wild Europe, giving up on purity necessitates a discursive shift from the "unspoiled" to the "untamed."[26] Here, the emphasis shifts from pure forms to lively (perhaps even vitalist) processes kept at bay by agriculture. They present rewilding as a "concept that does not aim at the fixed conservation of particular species, habitats or a priori lost landscapes, but rather opens for the continuous and spontaneous creation of habitats and spaces for species."[27] This enthusiasm for spontaneity raises a series of epistemological and political points that I address under the second and third axes.

Perhaps as a consequence of the environmental history of the country, Dutch conservationists seem much less attached to the preservation of a fixed cultural landscape than do their British and French colleagues.[28] There is a strong national pride in the Netherlands in modern water engineering, land reclamation, and planning.[29] In framing the significance of the OVP experiment for regional conservation, Vera and his collaborators talk of "nature development." They speak happily about engineering "new nature" with large herbivores.[30] Understood this way, OVP provides one means of moving beyond the paralyzing politics of paradox in which much modern environmentalism and its critics seem locked.[31] There never has been a singular Nature to which we can return or against which we can dispute the authenticity of a purported reconstruction. OVP offers an alternative to the stale found–made distinction about which such paradoxes depend. It offers a space for wildlife without the impossible geography of wilderness.

A related controversy flared over the legitimacy of experimenting with cattle and horses at OVP. As the wild ancestors of cattle and horses—the aurochs and the tarpan—are extinct, Vera selected "back-bred" animals with hardy natures and wild aesthetics as his surrogate bovine and equine grazers.[32] These animals were not found in the wild, nor did they arrive of their own accord. They were taken from zoos, were once domesticated, and have been confined within the small reserve.[33] If these herbivores had arrived at OVP of their own accord (like the beaver) or had never been formally domesticated (like the deer in the neighboring forests of the Veluwe), then research could perhaps have involved just observation. This would have been the "Serengeti behind the dykes" that advocates imagined, with no veterinary obligations and very limited license to intervene.

An alliance of animal welfare campaigners, hunters, politicians, and journalists have argued that the cattle and horses are fundamentally made, or at least "kept," animals that should be subject to the animal welfare associated with experiments in made spaces like laboratories, farms, and abattoirs. This debate centers on the welfare of the cattle and horses when food becomes scarce and some animals die of starvation. In 2005 the Dutch animal welfare organization (Dierenbescherming) launched a court case in which their lawyers argued that SBB lacked the required permit and called for an "end to the experiment" at OVP.[34] The judge found in favor of SBB, accepting their argument that they no longer exerted "factual power" over the animals.[35] Here, the absence or presence of a property relationship became the key determinant of the found–made status of animals. Through this rare revoking of animal property, the cattle and horses in OVP were classified as wild, and SSB was relieved of any legal responsibility.

In practice this situation became a public relations disaster for SBB. A compromise was reached that sought to reconcile and move beyond the conflicting demands of living with found and made animals. In it a wildlife ranger, armed with a rifle and a silencer, patrols the OVP, identifying and killing those animals whose bodily condition and behavior indicate that they will not survive the winter. This has been popularly termed population control with the "eye of the wolf." In theory the ranger must become wolf to choose his prey, drawing on ethology and the skills and knowledges of the field sciences. In practice, as so little is known about wild bovine and equine behavior (let alone their interactions with wolves), the scientific criteria used to assess the condition of individual cattle and horses are adapted from those used to judge the welfare of farm animals. These have been developed on farms and in animal behavior laboratories. A novel set of relations have emerged here that conjoin practices associated with found and made sites.

Drawing on the work of anthropologist Tim Ingold, Klaver et al. celebrate the dedomestication of large herbivores in the Netherlands as the replacement of relations of trust over relations of dominance. Here, the expertise of animals is valued, and organisms and landscapes are given more scope to determine their own futures. They term this a process of "controlled decontrolling of ecological controls."[36] This tentative compromise develops a mode of biopolitics that seeks to secure the future of the individual animal, the species it represents, and the wider ecology it helps compose, aware that the best interests

of all three do not always align. It offers a model for experiments in
wild forms of interspecies companionship that appreciate the "beastly
places" of nonhuman life.[37] These need not make recourse to the impossible spatiality of wilderness and ease the carceral confines of agricultural domestication. I discuss this model in more detail in this
book's conclusion.

Order–Surprise

One of the most striking differences between wildlife conservation
at OVP and in the Hebrides relates to how each conceives and seeks
to order emergent and unanticipated ecological properties (i.e., surprises). The equilibrium archetype of premodern agriculture offers an
orderly, linear ecology that can be known, predicted, and managed.
Hypotheses can be deduced and tested. Surprises are anomalous. This
epistemology informs the comprehensive assemblage, traced in the
previous chapter, for identifying, monitoring, researching, and nurturing various species and habitats. It has helped establish the legal
framework and territories for European conservation. It sets clear targets of species composition against which success can be audited. In
developing his alternative, nonlinear "theory of the cyclical turnover
of vegetations" with its dynamic "ecological reference" of the forest–
pasture landscape,[38] Vera has had to challenge the performative inertia
of this assemblage. He has had to develop alternative "experimental
systems" for testing his theory and generating new ecological knowledge. In principle Vera's theory allows for the deduction of concrete
hypotheses that can be tested through a field experiment (and a corollary assemblage to monitor, manage, and audit the alternative referent coming into existence).[39] He presents his theory this way in his
influential book.

   In practice there has been remarkably little prediction, monitoring,
or active management. Until very recently, there were no targets, no
models, and no explicit action plan. Partly, this absence is due to a
lack of interest in ecological science from the government agencies
that own and manage the site and a wariness about drawing attention
to the seasonal hardships of the large herbivores. More fundamentally,

in our discussions with site managers it was clear that they had a very different ethos toward field experiments.[40] OVP became famous as a source of surprises, and those interested in its ecology were keen to nurture and learn from its inadvertent ecological processes. Vera explained that this required him to cultivate a more speculative approach to science and management that aimed to grant a great deal of agency to certain nonhumans and the ecological and hydrological processes they performed. He argued that this had generated a range of surprising ecological events and new ecological knowledge that challenged the existing paradigm.

For example, the return of carrion has encouraged a pair of rare white-tailed eagles to nest (formally) below sea level, displaying unprecedented behaviors unanticipated by ornithologists. Similarly, the dedomestication of the large herbivores has generated animal behaviors at both an individual and a herd level that have not been previously witnessed by scientists in Europe. Cattle and horses at OVP display demographic structures, herd dynamics, and individual coping mechanisms that confound experts on their domestic kin.[41] Similar unanticipated events accompanied the reintroduction of wolves to Yellowstone. There, scientists observed that by reversing the "trophic cascade" that resulted from the wolves' extinction, they had reestablished an "ecology of fear" that shifted the grazing behaviors of their large herbivore prey, with important consequences for the dynamics of rare plant species.[42] The speculative, inductive, experimental systems established by wildness on reserves like OVP frame these sites as uncertain epistemic "wild things,"[43] capable of generating surprises and putting accepted knowledge at risk. Sympathetic conservation biologists, including important figures like William Sutherland, welcome this approach for offering increased "openness" in conservation management and for generating new knowledge.[44]

The challenges of such speculative wildlife science and management are perhaps most clearly displayed in the efforts of conservationists at OVP to comply with the Natura 2000 legislation. Natura 2000 is the centerpiece of the EU nature conservation policy and legislation for rational management encountered in the previous chapter.[45] It identifies a list of rare and/or threatened species and habitats that should be monitored, modeled, and managed. The dynamics of their populations become the accounting framework for deducing the success or failure of landscape management. It creates a network of protected areas, like the Special Protection Areas designated for the corncrake. Natura 2000 establishes a natural order founded on the compositional ideal of a premodern ecology.

OVP was legally designated under Natura 2000, as it accommodates a host of target species, including thirty-one priority birds. Conservationists at OVP are seeking to understand nonlinear ecological processes, not just species patterns. The annual dynamics of rare species have not been their primary monitoring concern. This has caused problems. In 1996 the population of rare spoonbills at OVP dropped from three hundred breeding pairs to zero. These figures spread panic among the external ornithologists who detected it. Accusations were made that the increase in foxes at OVP as a consequence of high levels of carrion, coupled with poor water-level management, had led to the collapse. There were calls for proactive bird-friendly management, including a change in stocking densities and hydrological regimes. Eventually, the population at OVP bounced back, and many of the displaced spoonbills were found to have moved out to colonize the wider network.

This event left SBB exposed. They had not predicted it, were not managing for it, and could not offer comprehensive data to account for it. The successive independent commissions created to address controversies over OVP site management have demanded that more be done to comply with Natura 2000. Calls have been made for an improved "statement of management objectives" and a "system of environmental monitoring," including "analysis and modeling to identify current processes, predict future trends and to set thresholds to acceptable change."[46] Much of this advice aims to bring OVP in line with the prevalent modes of rationality seen in the previous chapter. It seeks to circumvent conditions of uncertainty and rationalize the stochasticity that characterizes the current management regime. Such advice engages in a form of anticipatory governance that seeks to foreclose surprises that might result in politically and ecologically undesirable eventualities.

For its critics the fluid, speculative approach to science and management displayed at OVP poses a series of administrative and political challenges. On the other side of the Channel, a research report commissioned by English Nature examining the implications of the Vera hypothesis for UK conservation dismisses it as much on the grounds of its bureaucratic inoperability as on its scientific merit.[47] Nonlinear processes are anathema to the audit culture of British conservation.[48] Other ecologists have argued that a lifting of management prescriptions creates space for less-desirable emergent properties that might pose biosecurity risks—in the form of invasive species, dangerous predators, and zoonotic disease.[49] Letting go of a safely composed Nature might lead to economic risks and the local and global diminishing of biodiversity. Meanwhile, political opponents have offered criticisms that the speculative management regime exemplified by OVP is feared as a Trojan horse for dismantling hard-fought pieces of environmental legislation and subsidy that protect culturally significant ecologies and marginal political economies in upland and other geopolitically marginal regions. This is, they argue, an experiment whose ends have been framed in advance.[50]

As with tensions over the found–made status of OVP, the present settlement represents an uneasy compromise. Vera and his colleagues do enough to comply with the ordering requirements of Natura 2000 but resist efforts toward proactive management for designated species. In response to the criticisms of the second International Commission on the Management of the Oostvaardersplassen (ICMO), they published their first management plan for the OVP.[51] They have beefed up their monitoring programs to help anticipate, detect, and, thus, learn from future surprises. Reading these documents carefully and attending to how they articulate and seek to guide field practice, we can detect evidence of knowledge practices that can attune and respond to a mutable and emergent world. In Steve Hinchliffe and his coauthors' terms, there is evidence here of a speculative mode of "knowing around" (not a prescriptive "knowledge of") wildlife, informed by management documents that offer a set of loose "diagrams" for desired emergent futures.[52] There are similarities here with the textual version of the corncrake census methodology customized and performed by Craig in chapters 2 and 4.

Secluded–Wild

Unlike the crofting landscapes of the Hebrides, OVP is uninhabited and under sole public ownership. No one was displaced in its creation, and its conservation does not directly impinge upon any human livelihoods. In principle the politics of its management should be fairly straightforward. As a high-profile demonstration site for a new mode of conservation, however, it has attracted a great deal of attention. In contrast to the fairly open-ended deliberations with nonhumans that characterize the management of OVP or the largely consensual politics of corncrake conservation, SBB has been reluctant to engage with interested Dutch publics in resolving points of tension. As a result they were not able to prevent OVP from becoming a site of controversy—characterized by high-profile and frequently antagonistic debates among public officials, scientists, birdwatchers, farmers, and animal welfarists. In this final analytical section, I draw upon and develop Callon et al.'s criteria for evaluating the relationships between "secluded research" and "research in the wild" in order to critically examine the political processes through which OVP is governed. I aim to map its deficiencies and offer some lessons. Due to limited space, I focus on debates over the welfare of the large herbivores, which have posed the most fraught public relations challenges to SBB.

Public concern for the welfare of the large herbivores at the OVP first surfaced at the end of the 1990s. Visitors to the reserve could see emaciated animals starving to death, and images of their plight were soon circulating on television and (more recently) on the Internet. At this point the Dutch animal welfare organization initiated their campaign, which culminated in a court case. The Dutch government responded to this controversy by assembling the first ICMO, which was charged with examining the management of the site and advising the government minister on how it might be improved. They published a series of recommendations in their first report in 2006. During the harsh winter of 2009, the cattle and horses were again seen to be threatened by starvation. The controversy flared up once more, and the responsible minister (of agriculture, nature conservation, and food safety) was forced to answer questions about OVP in parliament. He felt compelled to make an emergency intervention to feed the starving animals. The ICMO was recalled and asked to evaluate SBB's performance and to make further recommendations on animal management. They published their second report in 2010, which summarizes the "governance situation" at OVP as follows:

> [The] management [of SBB] is not strongly driven by research and monitoring outcomes, little stakeholder engagement takes place in decision-making, little openness of the management to ongoing practices, all resulting in strong opposition of some societal groups against the current management strategy of the area.[53]

In short, the ICMO argues that SBB is not conducting a legitimate experiment. They first invoke the epistemological criteria used to evaluate secluded research, suggesting that SBB is failing to comply with the fundamental positivist requirement of future falsification and the full disclosure of data.[54] There has not been enough transparency in the data collection and publication to qualify this as a rigorous laboratory experiment. Furthermore, by not stating an explicit protocol for testing a hypothesis, the management regime cannot be held to account. Turning to the public dimensions of the OVP controversy, the ICMO then takes SBB to task for not carrying out the "stakeholder involvement" they explicitly advocated in their first report. SBB is presented as out of touch, unaccountable, and undemocratic. This is a damning critique. In Callon et al.'s terms, OVP is neither "secluded" enough to qualify as science nor "wild" enough to be democratic. The organization is caught in the middle and fails on two counts.

The substance of these criticisms is illustrated in the ongoing image wars relating to OVP. Here, various parties have contested how the landscape and its animals should be visualized and understood. Prior to the animal welfare controversy, SBB was happy to present OVP as a secluded laboratory: a private space where scientists could experiment out of sight and without deliberation. But OVP is in the suburbs, and much of the site can be overlooked from bordering dikes and roads, as well as a railway line. Ornithologists with binoculars and animal welfarists with film cameras have found ways of seeing and visualizing what is taking place. For example, the amateur films of animal starvation appear to have been shot at the perimeter of the OVP. Invoking the affective logic of footage smuggled out of laboratories and slaughterhouses, they show starving and dead herbivores interspersed with iconography of captivity and implicit references to concentration camps. Their grainy quality and shock aesthetic serve to heighten the sense of illicit practice and foreground claims of abnegated responsibilities.[55] I discuss this affective logic in more detail in chapter 6.

In response to these images and their political power, SBB and other rewilding advocates have changed tactics, promoting alternative visualities associated with field sites and nature reserves. For example, a photographer has been commissioned to produce a series of online videos exploring and explaining the OVP wildlife. His wildlife photography at the OVP has been published by SBB.[56] Targeted at the coffee tables of the Dutch middle class, it presents iconic plants and animals as denizens of the European wilderness. A feature-length wildlife documentary is currently in production, and access to the OVP via jeep safaris and bird hides has been promoted, including exclusive bookings for high-end private events. A different (and highly popular) strategy has involved deploying webcams to show the charismatic foxes of the OVP, evoking anthropomorphic features and familial narratives.[57] These visual rejoinders accentuate the found character of OVP as "the Serengeti behind the dikes."

While these images and practices constitute a form of public engagement, they continue to present OVP as a site that is accessed and known by a small cadre of scientists. There is a pressing need for further deliberation. In his work on coastal engineering in the Netherlands, Wiebe Bijker notes that the formation of state-sponsored scientific commissions (like the ICMOs) is a popular technique in the Dutch "technological culture" for deferring and/or delegating decision making to already-existing experts.[58] The ICMO was charged with finding a technical and managerial solution and consulting with key stakeholders but not with reflecting on the procedures by which decision making about the management of OVP could and should take place.

The successive commissions offer important first steps, but they are modest, conservative interventions. To use Callon et al.'s terminology, the ICMO is characteristic of a "delegative" model of democracy reliant on the "aggregation" of already-existing expertise to answer a preexisting question. There is little evidence here of their "dialogic" model of research in the wild in which the composition and expertise of a collective decision-making body emerges through the deliberative process.[59] Much of the ICMO's critique of SBB centers on their perceived failure to control the ways in which the management of OVP has been made public and visible, not with the openness of the management procedures themselves. This is perhaps most clearly conveyed in tactics for public engagement promoted by the ICMO and deployed by SBB. The focus here has been public education, employing various "experts in communications" to help frame the findings for external audiences.

While these attempts have gone some way toward persuading the Dutch public of the legitimacy of the experiment, the current approach is redolent of the "deficit model" of public understanding of science that has been heavily criticized in the sociology of science. There is a great deal of scope here for creating new "hybrid fora" to expand the deliberative ethos so far cultivated by SBB in its experiments with the large herbivores to include a more-than-human collective of sociable actors and emergent forms of expertise.

THE PROMISE OF WILDING

This chapter focuses on a case study of rewilding, an alternative paradigm for nature conservation that departs from the orthodox model of conservation as composition outlined in the previous chapter. I understand this paradigm to comprise a different mode of biopolitics, which I describe as engaged in a series of wild experiments. To explore the parameters, differences, and promises of this approach, I offer ways of specifying and differentiating expert-led experiments: first, according to where they take place and the difference these "locatory" properties make to their conduct and, second, by the relations they establish with the human and nonhuman constituencies they concern. Linking these differences, this chapter presents three of several possible axes for describing and evaluating real world experiments: found–made, order–surprise, and secluded–wild.

In conclusion, I flag some of the similarities, differences, and tensions between rewilding and conservation as composition. As part of this book's conclusion, I reflect on their combined promise. In relation to the Hebrides and the Dutch polders, I have been concerned principally with the epistemic properties of one specific locatory—the nature reserve. OVP and the RSPB's corncrake reserves are exemplary of these liminal spaces. Both are conceived as both "found" analogies for prehistorical or premodern pasts and as "made" sites for knowing and experimenting with different futures. OVP seeks to offer a model for rewilding Europe, whereas the corncrake reserves seek to show how this might be avoided and how agriculture might be protected. Both offer prestigious, visible locales for doing environmental science and for demonstrating forms of management; they are "heres" that advocates would like to replicate everywhere.

One seeks land sparing; the other, land sharing. The RSPB presents the Hebrides as a cultural landscape whose wildlife is dependent on low-intensity agriculture in inhabited landscapes. OVP is uninhabited and uncultivated, but it is not purified. The subterranean topography, artificial hydrology, and suburban geography are visible and acknowledged. Both offer a space for wildness without the impossible geography of wilderness. There are promising ontologies here for the Anthropocene, where to rework Callon et al.'s terminology, the wild is a multispecies commons. Where they differ most prominently is in how this wild is understood and governed. In the Hebrides the concern is with the population dynamics of species. At OVP the concern is with ecological, ethological, and evolutionary processes. These contrasting ontologies result in very different modes of science and management.

The deliberations with self-willed nonhumans at OVP align best with the second definition of an experiment outlined at the start of the chapter. Although the contemporary ecology of OVP is presented as a test of Vera's hypothesis, in practice it is valued for its ability to surprise. Freed from the management prescriptions associated with ensuring convergence toward a transcendent equilibrium Nature, OVP generates nonanalog events, behaviors, and ecologies. In contrast, science in the Hebrides adheres to a more traditional understanding of an experiment designed to test a hypothesis deduced from equilibrium ecological theory. The theory helps establish strict targets for averting undesirable change. In practice this is too neat a distinction. Both examples have elements of Rheinberger's experimental systems—arrangements of materials designed to generate and learn from difference—though much of that open-endedness gets written out of public accounts of corncrake science. Taken as ideal types, they help indicate the difference.

What is taking place at OVP would therefore seem to have a great deal to offer environmentalism in the Anthropocene given the current uncertainties about the nonlinear ecological responses to accelerated climate change and the conditions of uncertainty that currently characterize research and planning for climate change adaptation. But environments cast off from a fixed Nature and operating in the wild outside the laboratory (or equivalent computer models) are much more politically risky than those in the Hebrides. Nonequilibrium ecology offers few universal criteria for identifying failure or for specifying undesirable future scenarios, however self-willed. OVP gives a good illustration of the risks of advocating a shift to immanence.

Many of the local opponents to what is happening at OVP are defending clearly specified natures, like those associated with animal welfare, the future of rare birds, and the demise of the cultural landscapes they inhabit. These are familiar and commendable political projects with hard-fought territorial and legislative gains. There is a real risk that rewilding, with its purported open-ended ecology of surprises, could inadvertently play into the hands of those who would like to see them removed. As James Evans notes, fungible, laissez-faire neoliberal natures and fluid, self-willed ecologies are ontologically not that different.[60] As I discuss in more detail in the conclusion, it is vital that in conceiving of the wild, in which the wild experiments at OVP take place, we keep sight of a set of wider debates about the current and future political ecology of Europe that will frame how these experiments proceed and to what ends they are put. Clearly, we need to be cautious here.

· 6 ·

WILDLIFE ON SCREEN

The Affective Logics and Micropolitics of Elephant Imagery

We live in an age of the screen, surrounded and enveloped by moving imagery. Contemporary nature conservation takes place in and depends upon this mediated ecology, where Western publics are much more likely to encounter the charismatic organisms about which they are so concerned in print, online, or on TV than they are to meet them in the flesh. Elephants, tigers, and polar bears are now quotidian features of media landscapes, their representations proliferating even as their fleshy kin and vital ecologies disappear.[1] Watching animals is a popular and lucrative source of entertainment, and evocations of wildlife have governmental effects; they come to shape how people sense, think, feel, and act toward the nonhuman world.[2] Wildlife filmmakers frequently justify their activities on such pedagogical grounds, arguing that the entertaining spectacle of nature will catalyze desirable environmental subjectivities.[3] Campaigning conservation and animal welfare organizations mobilize charismatic organisms to pull at the heartstrings and open the wallets of watching publics. These mediations aim to shape the worlds they purport to represent and thus have biopolitical implications.

In this chapter I examine the role of moving imagery in nature conservation. While visualization is an important tool for conservation science, here I focus on popular and artistic imagery. I give an overview of the ways in which wildlife is evoked. Not all of these are directed explicitly toward conservation, but together they contextualize how conservation might be conducted. The content and power of wildlife media have become familiar concerns for social scientists interested in the representation, commodification, and governmental power of environmental knowledge. There are copious literatures on the history, politics, and economics of nature writing, photography, campaigning, and natural history film.[4] Jonathon Burt argues that these analyses are concerned largely with "textual animals," which are understood to be impoverished or fetishized representations of "real" animals and their human relations.[5] These are then subjected to critical analysis to unpack their discursive and political–economic power. This approach has produced a rich and important body of research that has revealed diverse ideological undercurrents in, as well as political economies of, animal representation.

This work pays limited attention to the affective properties of media ecologies—especially the ways in which "electric animals"[6] are mobilized to cue strong emotional responses (often in support of the ideological commitments that concern the textual animal detectives). Here, I offer a more-than-representational account that supplements existing work by attending to the affective dimensions of wildlife media. Building from the introduction to affect that I offer in chapter 2, I identify four prevalent and distinctive affective logics to wildlife film. Here, filmmakers work with the aesthetic charisma of particular nonhumans to move their audiences in diverse ways and toward particular ends. I first offer a brief introduction to the theoretical foundations of this approach before returning to elephants. Working through popular evocations of elephants, I illustrate and critically examine the techniques through which my four logics work. I reflect on the types of environmental subject and modes of human–nonhuman interaction each summons forth.

I should be clear from the outset that I do not share the common sensibility of many conservationists that the proliferation of wildlife media is complicit with the "extinction of experience" and the spread of a "nature-deficit disorder."[7] This popular argument contends that authentic encounters with Nature "out there" are being replaced with alienated, sedentary encounters at home, with detrimental effects on the future health of environmental citizens and support for conservation. I offer a more nuanced account. It examines how certain evocations of wildlife on screen perform environmentalities with detrimental consequences for the biopolitics of conservation. This position is counterbalanced by a more affirmative critique that nurtures the potential of diverse wildlife media to inculcate a curious sensibility toward nonhuman difference that I believe necessary for conservation after the Anthropocene.

AFFECT, FILM, AND BIOPOLITICS

Theorists interested in life in the age of the screen have argued that moving imagery does not transmit visual representations through the eyes solely to a cognitive mind. They suggest that attending to the material, practical, and affective dimensions of moving imagery enhances critical understandings of their power and provenance. Alan Latham and Derek McCormack argue that "the force of images is not just representational. Images are also blocks of sensation with an affective intensity: they make sense not just because we take time to figure out what they signify, but also because their pre-signifying affective materiality is felt in bodies."[8]

Film theorists argue that we need to understand moving images as rich, mimetic, multisensory, and affective media. For example, Donato Totaro proposes that we should understand cinema (and by extension most other modes of moving imagery) as "the ultimate synesthetic art, incorporating sound, voice, music, color, movement, narrative, mimesis and collage in a fashion so visceral and emotive that it can frequently move spectators to think and feel beyond the sensorial limits of sight and sound."[9] For writers like Laura Marks and Vivian Sobchack, it is the "haptic visuality" of moving images that give them their allure and evocative power.[10] In Donna Haraway's terms, moving images create "fingery eyes," performing "heterogeneous infoldings of the flesh" that trigger embodied senses of "response-ability."[11]

This understanding of moving images chimes with or is indebted to Gilles Deleuze's revolutionary rethinking of cinema through the philosophy of Henri Bergson.[12] Deleuze develops a philosophy of cinema that maps a taxonomy of "images" understood as "resonant blocks of space-time"[13] and explores how images are strategically combined and juxtaposed to evoke form, affect, and movement. Deleuze provides means of understanding and evaluating moving images that dispense with critiques relating to signification, representation, and desire in order to explore how films can provide a sensual "shock to thought," catalyzing, but more often restricting, ways of feeling with concomitant effects on governing life.

The broad argument across this work is that moving imagery frames our "optical unconscious," organizing the horizons of the visible and the sensible.[14] One compelling articulation of this approach can be found in the work of Kathryn Yusoff, who draws on Michel Foucault's and Jacque Rancière's writings on "political aesthetics" to examine how "the distribution of the sense experience is crucial to the political spaces of biopolitics."[15] She explains how for Rancière the space of political aesthetics is "the system of *a priori* forms determining what presents itself to sense experience. It is a delimitation of spaces and times, of the visible and the invisible, of speech and noise, that simultaneously determines the place and stakes of politics as a form of experience."[16] This thinking informs her critical analysis of how the aesthetics of environmental media comes to frame possible responses to climate change and biodiversity loss.

The critical analysis of moving images in this vein examines the techniques employed to relate image, sound, and narrative in styles that evoke and mobilize particular "affective logics" toward various political ends.[17] As explained in chapter 2, I understand an affective logic to describe a particular embodied disposition that establishes a habituated set of practices and feelings, often occurring in advance of reflexive thought, through which a person orients himself or herself within and makes sense of an encounter with human and nonhuman others. The argument in this work is that affect, understood in this fashion, has become the target of powerful modes of biopolitics.[18] These make ready use of moving imagery. Recent work on film has explored the geopolitics of fear and nationalist anxieties around security.[19] To date there has been very little work exploring the affective logics of wildlife film.

For political theorists like William Connolly and Brian Massumi, moving images should be understood as important "neuropolitical" mechanisms through which "cultural life mixes into the composition of body-brain processes" to frame thought and action.[20] Connolly argues that film techniques mix sound, image, words and rhythm together to work on the visceral register of human sensibility. It is the intersection between techniques and story which is critical. Attention to such intersections discloses how immersed we are in the sea of micropolitics. By micropolitics I mean, for starters, organized combinations of sound, gesture, word, movement and posture through which affectively imbued dispositions, desires and judgments become synthesized.[21]

Connolly suggests that moving imagery and other media are vital components of affective "resonance machines" that powerfully configure popular political landscapes.[22] Jane Bennett and Michael Shapiro argue that this approach "foregrounds the connections between affective registers of experience and collective identities and practices" and provides the foundation for a positive micropolitics whose

aim is to encourage a more intentional project of reforming, refining, intensifying, or disciplining the emotions, aesthetic impulses, urges and moods that enter one's political programs, party affiliations, ideological commitments and policy preferences. . . . The claim is that politics in the broadest sense . . . requires not only intellectual codes . . . but also an embodied sensibility that organizes affects into a style and generates the impetus to enact the principles, programs and visions—or to reveal the singularities they exclude.[23]

This radical rethinking of the relationship between moving imagery, affect, and politics informs emerging styles of "affirmative critique" that seek to reinvigorate left and environmental politics.[24] As I hope to demonstrate in the analysis that follows, the stance developed by these authors provides powerful resources for engaging with evocations of wildlife in moving imagery and for interrogating the wider role of media in the biopolitics of nature conservation.

EVOKING ELEPHANTS

Elephants provide an excellent example of the virtual electric animals that flourish in contemporary media ecologies. They are charismatic and telegenic and have proliferated globally even as their fleshy kin decline. They are mobilized for diverse purposes—some directly connected to the bodies, ecologies, and fates of their threatened progenitors, others more closely linked to art, commerce, and entertainment. In all of these cases, different elephants are strategically evoked in media presentations that catalyze different affective logics. Each of these presentations has material effects of varying nature and magnitude on the animals themselves. My aim is to map four logics, the model of the moving animal they present, and the more-than-human "micropolitics" they perform.[25] I have termed these *sentimentality, sympathy, awe,* and *curiosity.* For each logic I explore the practical techniques employed by image makers to elicit affection and examine to what ends these are put.[26] For analytical clarity I illustrate each logic with images drawn primarily from a particular genre: animation, natural history documentaries, campaigning film, and experimental media, respectively. This typology is heuristic. In practice, work in each genre evokes a variety of affects, and no logic is essential to a genre. The ultimate aim here is to nurture the potential of moving images to open thinking and feeling spaces for the mobile, mutable, and emotional dimensions of difference (in this case elephants) and thus push for different, more convivial political/ethical sensibilities toward (non)human others.

Sentimentality

Animals are a staple of popular animation, where they appear in anthropomorphic forms and allegorical narratives to play on and affirm familiar human emotions. Diverse affective logics are at work in this genre, but one that predominates is sentimentality. Here, moving images are used to trigger and develop basic and often clichéd feelings in the audience. Love, pity, anger, and humor are evoked in fantastical styles that drift toward mawkish nostalgia. Such renditions are powerful and are achieved through a repertoire of filmic techniques that have been developed over the long and distinguished history of animation. Elephants are popular protagonists in animal cartoons, featuring luminaries like Elmer, Babar, and, most famously, Dumbo.

Dumbo is an award-winning animated feature film produced by the Walt Disney Company in 1941. It is perhaps the simplest and most moving of Disney's animations. It tells the story of an unfortunate baby circus elephant whose large ears and clumsy demeanor initially lead to his rejection from the troop but ultimately result in his salvation when he learns to fly. Its affective force is created by at least three interwoven sets of techniques. The first is the universality of its story. We identify with the main character and his relationships, which are framed within ubiquitous themes: here we have a bullied outcast separated from his mother who finds acceptance and success through heroic endeavors. Our affections are clearly directed from the start and are conducted by sophisticated narration and shot sequencing.

Second, it is the quality, credibility, and evocative power of the animation that make the film work. Disney filmmakers did not aim to provide a realistic representation but sought to caricature, reducing the complexity of an animal and accentuating the features and behaviors that they expected to be most affecting to their audience. This is perhaps best conveyed in their evocation of the cuddly charisma of infant mammals, of which Dumbo is an archetype. Indeed, *Dumbo* is the only Disney feature film in which the lead character doesn't talk. All of his characterization is mimetic, and the powerful associations we form with him are created and sustained through the careful use of close-ups—what Deleuze terms "affection-images," which frame and mobilize communicative parts of a subject's anatomy.[27]

Dumbo accentuates the elephants' weeping and gazing eyes (and uncommonly expressive eyebrows) and his prehensile, touching trunk. For example, during the famous "Baby Mine" sequence, where Dumbo is reunited with his mother, we first see his drooping face and shuttered weeping. On his way to his mother, his eyes open and brows rise, and we sense his anticipation. Then having touched and entwined their trunks, his eyes glaze over in happiness as he is cradled and rocked. Here, Dumbo's eyes provide a window onto a sentient and suffering soul.[28] *Dumbo* foregrounds the human anatomical features, discussed in chapter 2, that trigger human affection and ethical concern—the eyes, face, and hands.

Third, *Dumbo* is a particularly stylized cartoon that works with a gaudy logic of sensation. The animators used broad brushstrokes and great washes of color to suggest mood. At times this is riotous and uplifting—with imagery steeped in the visual vernacular of the circus (all primary colors)—and at other moments the gloomy light and inclement weather combine pathetically to mirror Dumbo's dejection. These visual effects are accentuated by the pastiche musical score, which ranges wildly from weeping strings to triumphant horns, punctuated by elephant trumpets and the full syntax of cartoon sonic punctuation—*crash, bang, wallop,* and the rest. This brash and familiar sonic landscape provides the glue that holds the narrative together and propels us through its emotional peaks and troughs.

Disney's animations are incredibly successful at evoking sentimentality. They work off a lucrative formula that guarantees tearjerking, heartwarming, and teeth-grinding moments to sympathetic audiences. Childhoods steeped in such viewings no doubt influence citizens' sensibilities toward charismatic animals in later life. Indeed, it could be argued that efforts to conserve species such as elephants, pandas, and tigers would have gotten nowhere without the moving images of Disney and his colleagues in the middle of the twentieth century.[29] Sentimental appeals exert a powerful influence on the wallets and politics of the urban middle classes. Alan Bryman and Alan Beardsworth note how Disney films also come to script expectations of and encounters between people and captive animals at zoos and Disney theme parks.[30]

Sentimentality has its critics, who can be found at different points on the political spectrum.[31] Some of the most trenchant of these are in avant-garde poststructuralist philosophy, for whom these safe and lucrative renditions of individual animals serve merely to affirm an anthropomorphic affective landscape and tread a familiar narrative path full of cliché. The micropolitics they perform is conservative; it involves repetition and the continued disciplining of affect according to an order of the same. For Deleuze and Guattari, evocations do not provide a shock to thought, nor do they do justice to the living difference of animals and their ecologies.[32]

Many of these criticisms can be applied to *Dumbo*. Disney's animators domesticate the elephant with their caricatured anatomy. With extending ears, curling tails, fattening bodies, and shortening jaws, they give Dumbo and his fellow elephants many of the features that archaeologists associate with organisms subject to generations of selective breeding.[33] This cartoon "petishism"[34] succeeds where thousands of mahouts have failed. In Deleuze's terms, much of this film is classic "movement-image" cinema in which the syntax and narrative of the imagery propel the viewer smoothly through a familiar if dramatic affective landscape toward an expected conclusion.[35] There is little that surprises in this film, which was produced in 1941 against a backdrop of global rearmament and political upheaval for an American public seeking comfort in the sentimental nostalgia of animated elephants. The one notable exception is the famous pink elephants on parade scene, in which an inadvertently drunk Dumbo hallucinates surreal, shape-shifting, and transgressive kinfolk who drink, dance, and make merry.[36] Although this sequence runs counter to the conservative logic of the main narrative, the psychedelic elephants it features have become yet more human, standing upright like Orwell's pigs to skate, jive, and play musical instruments. In short, even in its most surreal moments, *Dumbo* as allegory reduces the alterity of emotional, living elephants to an anthropoidentity.

Sympathy

The affective logic of sympathy is concerned with the lived experiences of real, individual living animals and the humans they encounter. It seeks to establish close, familial, and loving affections between the audience and the target animal and to mobilize these affections to shock in the context of animal cruelty. Sympathy and sentimentality are similar but can be differentiated by the greater scope afforded by sympathy to the alterity of the subject of care. Sympathy can be extended to an individual nonhuman other without reducing that other to a subhuman form. This logic of sympathy toward elephants is expressed most clearly in one strand of recent documentary filmmaking, as well as in the campaigning imagery of animal welfare organizations protesting against the treatment of circus elephants.

A logic of sympathy in elephant documentary characterizes the work of Martyn Colbeck, who directed and filmed the BBC's three-part series *Echo and Other Elephants,* among several others.[37] These films follow the lives of Echo, a matriarch African elephant, and other individuals from her multigenerational family in Amboseli National

Park in Kenya. They are informed by the extensive expertise and caring ethos of the elephant ethologist Cynthia Moss and are centered on "pachyderm personalities" as they are understood by Moss and Colbeck.[38] A logic of sympathy also characterizes the genre of presenter-led wildlife documentary that focuses on the rehabilitation of individual animals in captive settings. Orphaned African elephants at a sanctuary in Kenya were the recipients of such televised celebrity care in the BBC series *Elephant Diaries*.[39] This approach breaks down the aesthetic of distance that is commonly associated with wildlife documentaries and bestows celebrity status upon individual animals.

To create this affective logic of sympathy, Colbeck employs many of the techniques I identify as at work in *Dumbo*. Individually and together, the *Echo* films have a coherent narrative. Their success relies on us becoming familiar with the presented pachyderm personalities and the universal mammalian stages and events they experience. The central narrative conceit is the role of the matriarch and her relationships with her children. During the fifteen years the programs cover, we are skillfully conducted through a dramatic range of emotions common to both humans and elephants. We witness tragedy, pain, joy, play, bravery, and grief, and these movements are framed around familiar rhythms and events. We learn of the seasons, of cycles and generations and are treated to disasters, novelty, and spectacles.

As in *Dumbo*, sympathetic affections are triggered, and the momentum of the films are maintained by the careful montage of images, sounds, and music. Great use is made of close-up affection images. At moments of high emotional intensity, the image flow repeatedly cuts to the elephants' eyes (rather confusingly, as without Dumbo's dynamic eyebrows, they are fairly inexpressive organs). Attention is paid more effectively to the touching and lithesome intertwining of trunks. A samba score is keyed to evoke drama and to signal play, while emphatic trumpets and the deep sonic rumble of elephant vocalizations add a mysterious yet evocative sound track.

The *Echo* films are overwhelmingly movement-image television in which temporality and plot are ordered and the narrative reaches for meaning and resolution. The director is modest, however, and uncertain enough to acknowledge the unknown and to hint at elephant difference. This is achieved (perhaps inadvertently) through the use of elephant calls that often transgress the narrative: supposed shrieks of pleasure sound like pain, while distant seismic rumbles interfere, making the hair stand up on the back of your neck. Similarly, the continual cuts to the elephant eyes are disconcerting; we cannot be sure that they express what David Attenborough's commentary wants us to believe. There is a gap, an aporia in the tight interspecies attunement supposedly on display. It is at these points that these films are most experimental and thought provoking. With this dissonance they begin to evoke dimensions of elephant difference effaced by the clichéd, beguiling logic of *Dumbo*.

Sympathy is mobilized more dramatically in images produced by animal rights organizations, which aim to both inform and shock audiences about the treatment of animals in captivity. The online TV channel of People for the Ethical Treatment of Animals (PETA) hosts a wealth of such campaigning materials, including several films documenting cruelty to circus and tourist elephants.[40] These films and their accompanying materials are didactic and moralistic and have a strong political message. Their discursive impact depends in part on the force of their argument, but it is also reliant on the successful deployment of a range of affective techniques.

For example, PETA produced a short film as part of its campaign against the use of captive elephants in circuses in North America.[41] It comprises footage shot by amateur or undercover observers or taken from news bulletins. In contrast to the professional camera work and slick editing of Disney or the BBC, it aims for gritty (and often grainy) realism. This type of low-resolution footage and haphazard amateur editing is ubiquitous in the age of video cameras on mobile phones, YouTube, and viral emails. It is rarely given the oxygen of prime-time TV. As such, it has an illicit feel, which is employed strategically to make us believe that these are shady practices happening in hidden places. Here, the camera takes us where we would not or could not go, revealing spaces, bodies, and events generally obscured from contemporary visual horizons. Such images are primed to erupt spectacularly into public view, courting controversy and reaction.

The majority of the footage in this film is composed of lingering mid- and long shots taken from a standing point of view. There are few cuts, and viewers find themselves in the scene as distant observers.

The poor quality of the photographic equipment and the need to stay away from the elephants prevent the recording of affective close-ups. Without this weapon of film syntax, we are not given the same sense of elephant emotion and do not identify with individual animals. The film works hard, however, to create mimetic triggers that engender sympathy and outrage. Images of rampaging elephants being shot are sensational and shocking and would seem to demand a response.

In differing ways, these films employ electric animals to open sympathetic spaces of sensation that provide a shock to thought. In the *Echo* trilogy, this is done subtly. Colbeck's films create Hayward's "fingery-eyes,"[42] using technology, imagery, and sound to bring the lives of elephants to our screens and allowing momentary connections in which we get a sense for the mysterious life of other beings.[43] PETA's interventions are more blunt and visceral. They target the gut instincts of repulsion and disgust, trusting that an exposure to such imagery will make it impossible to sit idly by. In comparing the micropolitical efficacy of these two approaches, it is clear that brute shock can be effective but is difficult to sustain. Such radical, moralized encounters can lead to exhaustion, apathy, and even cynicism. Many of the PETA images appear fantastical to cosseted eyes and, in keeping with much of the amateur animal imagery to be found online, feel both voyeuristic and horrific. My sense is that shock is too uncomfortable an emotion upon which to base a micropolitics; it is reactive and requires an active, positive corollary. Colbeck's subtle and curious sympathy would appear to offer a more sustainable foundation on which to base a critical environmental politics, but its pleasant palatability makes it all too easy to consume—there is little that niggles after the credits have rolled.

Awe

Elephants have been popular subjects for documentary film from its inception. Celluloid elephants abound in the archives of early moving imagery. These are littered with short clips of performing circus elephants and of elephants being hunted and ridden and on procession in colonial Africa and South Asia. Perhaps the most notorious is Thomas Edison's film of an elephant being electrocuted. It was created to promote the power of Edison's invention. As Rosemary Collard observes, it also serves as a useful reminder of the unrecognized ecological damage and animal cruelty wrought by the film industry.[44] Most of these early images were shot at the turn of the twentieth century as film was being invented. They show little interest in the animal itself but feature elephants (alongside other beguiling or freakish phenomena) as large, lively, and charismatic beings that illustrate the affective potential of film to help draw in larger audiences.

The desire to witness and present the "real" behavior and ecology of animals emerges later, in the 1930s and 1940s, with the pioneering work of ethologists and conservationists.[45] Their early wildlife documentaries adopt a curious sensibility toward life in the nonhuman world. They use moving images to research behavior and ecology and to educate their audiences in the insights of their new science. I discuss this logic of curiosity in the following section. Subsequent filmmakers translated the didactic tone, erudite language, and visual and affective vernacular of zoology into moving imagery, employing what Greg Mitman terms a "calculating aesthetic of distance." This stands in stark contrast to the sentimentality and anthropomorphism of Disney's animations.[46]

In its popular manifestations in contemporary wildlife documentary, curiosity and calculation seem to have given way to awe and the affective logic of the sublime. The aim here is to evoke the overwhelming size, power, and alterity of nature to provoke admiration, reverence, and fear. In relation to elephants and their ecology, this logic is perhaps best expressed in *The Life of Mammals,* a series of programs that feature the animals from the BBC's flagship *Natural World* nature documentary strand. Here, Attenborough's hushed, patrician tones narrate a story of ecological and evolutionary processes rather than of animal affections and familial bonds. In stark contrast to Colbeck's work, the featured elephants are anonymous, almost incidental. Individuation and audience identification are discouraged by the frequent use of aerial photography, flyovers, and sweeping panoramas of depopulated and objectified landscapes. Close-ups illustrate behaviors, not emotions.

Wildlife films with this logic invest heavily in sensation. Aided by bigger budgets and technological developments (like high-definition video, Imax, 3-D, and Crittercams), they increase the emotional intensity of their presentation to evoke awe and drama. Sophisticated nonlinear editing packages allow filmmakers to compress time and space and manipulate sound to accentuate behaviors and events and thus create tension and excitement. The narrative voice goes base, and the tone becomes epic. The score is orchestral, full of soaring wind and crashing percussion. New or digitally enhanced sound effects complete the effect in postproduction. For proboscideans this dramatic, awe-full logic reaches its apotheosis in episode six of the BBC's CGI-driven series *Walking with Beasts*. Here, digitally created mammoths battle ice age conditions, Neanderthals, and cave lions as they migrate, graze, and fight their way across the snow-covered plains of western Europe in the Pleistocene.[47]

The moving animals evoked in this register are fundamentally wild and different. Great attention is given to portraying their alien ecologies, unfamiliar anatomies, and inhuman behaviors and, in so doing, to eschewing the anthropoidentities that characterize the logic of sentimentality. The logic of awe on display here and the nonhumans it evokes have a long cultural tradition. They are linked to the romantic conception of the sublime. In popular incarnations—especially those targeted at North American audiences—the sublime elides with the cult of wilderness and an apocalyptic understanding of environmental change. Here, wilderness is a hostile place without people and the cradle of extreme forms of political and economic individualism.[48] Nature is an avenging angel hell-bent on dystopic destruction. Recent work on "inhuman nature" has sought to rescue the sublime for the Anthropocene, acknowledging the affective power of volcanoes, tsunamis, hurricanes, and other disasters to generate respect and catalyze human hospitality and generosity.[49]

The sincerity of the respect for nonhuman difference summoned forth in much awe-some wildlife imagery is often open to question. There is a tendency to drift toward the pornographic in these evocations—we are presented with an improbable feast of expansive and unpopulated locations inhabited by exotic animals, which are forever fighting, fucking, eating, migrating, and dying for their impatient channel-surfing audiences. Rooted in the linear temporal logic of the movement-image, these images seek closure in the presentation of gory excess and a romantic affirmation of a pure and thrusting nature.

Political ecologists have argued that the inhabited spaces imagined in these images—for example, in the BBC's recent natural history series *Africa*—naturalize the histories of violence and expropriation that led to the contemporary depopulation.[50] Drawing on Marxist critiques of commodification, they argue that these images act as spectacle—visual fetishes that enable naïve and alienated consumption.[51] This works in the nominal interest of wilderness conservation while masking the socially and ecologically disastrous relations of their production. Such images prevent citizens from establishing true and authentic relations with local, mundane wildlife. I engage this critique in more detail in the following chapter.

Curiosity

The disconcerting force of moving imagery is more explicitly and radically channeled in a final genre, which can loosely be termed *experimental media*. It encompasses a diversity of fields, including surrealist wildlife documentary, postmodern animal art, and experimental video, all of which share a desire to disconcert the viewer, using the haptic visuality of moving imagery to provide shocks to thought, challenging the syntax of orthodox imagery and its associated animal evocations, and inventing new techniques for imagining animal life and human–animal interactions differently. I describe this as a logic of curiosity. It differs from the logics of awe and sympathy by virtue of the human–nonhuman relations it concerns and the style in which these are evoked. The subject of curiosity need not be a sentient animal body or a purified, powerful ecology. More often, it focuses on humans and nonhuman difference in their intimate entanglements in bodies and practices.

There is a rich minor tradition of experimental filmmaking in French wildlife documentary that can be traced back to pioneers like the surrealist Jean Painlevé. Painlevé's films (largely produced between 1929 and 1965) set the balletic movement of marine organisms to experimental electronic music.[52] Here, the narrative is sparse, discontinuous, and irreverent. The settings are wonderful and alien, and the score is jaunty and jarring, avant-garde in its obvious presence. Painlevé's approach influenced Jacques Cousteau, and it is clearly echoed in more recent work by the directors Claude Nuridsany and Marie Pérennou.[53] These films explore interspecies commonality and difference, employing what Jim Knox terms an "elliptical anthropomorphism." They aim to unsettle, educate, and provoke curiosity by revealing unsentimental, absurd, violent, and erotic universals that cut across species and spaces.[54] In a detailed analysis of Jean Painlevé and Geneviève Hamon's *The Love Life of an Octopus,* Eva Hayward draws on Haraway and Marks to explain how

> the film mobilizes an alternate economy of visceral *seeings* that enfold the spectator into an "intra-acting" relationship of "empathic non-understanding" and/or "significant otherness." The film's refracted modality, through its convergence and torqueing of science, cinema, and surrealist practices, posits an ethics of interspecies relations that disrupt anthropocentric hierarchies.[55]

More mainstream incarnations of this playful aesthetic of interspecies intimacy, connection, and dissonance occur with some regularity in the BBC's long-running live outside broadcast series, currently incarnated as *Springwatch* and *Autumnwatch.* These events deploy a network of manned, womanned, and remotely operated cameras hidden in nature reserves to track the British seasons. While the format is grooved toward charismatic animals and spectacular events, mixing the logics of awe and sympathy, the wildlife and the vernacular audience frequently go off script, disconcerting presenters and in-house naturalists and forcing them to respond.

In his comprehensive reviews of the *Postmodern Animal* and the *Artist/Animal,* the art historian Steve Baker identifies the prevalence of similar techniques for disconcertion in the moving animal imagery of a number of contemporary multimedia artists, including Mark Dion, Bill Viola, and Joseph Bueys.[56] In some of their experimental, nonrepresentational work, these artists embrace the potential of what Deleuze terms "time-image cinema," where standard linear narratives are confused and left open. The aim is to make time present, asynchronous, and discordant.[57] The affective force of the imagery is emergent and underdetermined. Seemingly chaotic sequences of images depicting animals and humans, together and apart, are layered over each other, repeated and reversed to catalyze comic, confusing, and disturbing affects. With differing degrees of normativity, these works play with and challenge the clichéd affective logics I review. For example, in a video entitled *I Do Not Know What It Is I Am Like,* Viola plays with representations of animal eyes to explore human–animal encounters and questions of consciousness, making himself present in the image as a reflection in an owl's eye.[58] Images such as these seek to provide disconcerting mimetic resources for rethinking and appreciating nonhuman difference.

Elephants are poorly represented in this genre, perhaps for the very reasons that make them so popular in images informed by affective logics of sentimentality and sympathy. Experimental, avant-garde artists and filmmakers have tended to shy away from anthropomorphic animals and have instead focused either on the wild, the alien, the abject, and the out of place—on disconcerting animals performing various challenging modes of radical alterity—or on unlikely and enchanting cross-species encounters with mundane nonhumans in unromantic, quotidian spaces. The exceptions here relate to the symbolic use of elephants as visual incarnations of popular metaphors in which they feature—for example, the UK graffiti artist Banksy's *Elephant in the Room*[59] or Javier Tellez's *Letter on the Blind, For the Use of Those Who See,*[60] which reworks the famous Hindu parable of the six blind men and an elephant. This general absence of experimental elephants from this later genre is to be expected, but the lack of monstrous and outlandish moving elephants is perhaps surprising given the preference expressed for the animals by surrealist painters like Dali and the success of recent urban elephant installations, like *The Sultan's Elephant,* designed to enchant, educate, and provoke curiosity.[61]

The micropolitics of disconcertion expressed in experimental media operate in different registers to the cloying sentimentality, sympathetic outrage, or awe-full respect of the three previous genres. At its most deconstructive and critical, this work is steeped in postmodern irony and cynicism, concerned with either challenging modern divides or ridiculing obsessions with the cute and cuddly. In its more affirmative incarnations (that eschew the romantic preoccupation with the wild and the sublime), it attends to mundane nonhumans and forms of practical, cosmopolitan companionship. They are interested in alliances with difference without smothering them with care.[62] Moving images forged in this model open space for the emergence of unexpected affections and connections. There is an interest here in cultivating what Jane Bennett terms the "enchantment of modern life."[63] Invoking Spinoza, they suggest that we do not know what a moving image can do and set out to explore this potential in a playful style that is more open-ended than the three other genres. Given the absence of surreal moving elephants, it is difficult to judge and compare the political potential of this logic of disconcertion, though the colorful political history of experimental media hints at some interesting and fertile possibilities.[64]

VISUALIZING DIFFERENCE

This chapter focuses on the role of moving imagery in nature conservation. It argues that we live in the age of the screen, in which media assemblages have come to play vital roles. The Western public, upon whom conservation depends, are much more likely to encounter charismatic animals on screen than in the flesh. These encounters come to frame how people think and act toward wildlife. Here, I argue, though perhaps not altogether compellingly illustrate, that these encounters have governmental effects. The ubiquity of normative wildlife film forge forms of environmental subjectivity. Preferences cultivated and given expression in these media may well help perform the partial biopolitics of nature conservation outlined in chapter 2. Perhaps most significant, these media are frequently commodified. They serve to generate and channel financial resources, sometimes in the interests of the wildlife they represent. As I explore in more detail in the following chapter, such commodified media assemblages perform unprecedented and unequal geographies of international connection, bringing people and wildlife into novel political and economic relations.

In this chapter I develop a more-than-representational account of wildlife media. This account differs from those concerned with the representation, commodification, and performance of environmental knowledge in the attention it affords affect. Here, I examine the affective logics that characterize popular and artists' evocations of wildlife (in the form of elephants). I explore how these logics conjoin visual, cognitive representations of animal with the haptic visuality of film. I draw attention to the significance of sound and music, montage and composition, color and contrast, rhythm and discord in configuring the affective force of images felt through our bodies. This approach helps identify the techniques employed by moving-image makers to evoke these logics and offers new supplementary means for categorizing and critically interpreting existing imagery. There is much more work to be done here to unpack the relationships between moving imagery and affect in order to explore how images amplify or undermine the power of dominant discourses and ideologies.

Working from an analysis of evocations of elephants, this chapter explores the micropolitics of moving imagery with the intention of developing a critically affirmative vocabulary for making sense of and directing evocations of nonhuman difference. The micropolitics of elephant evocation is richly affective and is characterized by a complex topography of responses. Broadly speaking, my analysis identifies two contrasting tendencies in the affective micropolitics of elephant evocation, which emerge from the types of aesthetic charisma identified in chapter 2. At its extreme the first seeks to engender sympathy for individual animals by reducing them to anthropoidentities. Through the strategic use of comic and tragic clichés and resolved narratives, we are taught to be affected by animals that are just like us. The second, countervailing tendency seeks to inspire awe and respect through the pornographic presentation of extremes of difference. Here, we have sensational animals performing in wild spaces. Although they are beguiling and lucrative, the affective, mimetic resources made available by either of these extremes offer limited means for a multispecies politics for wildlife—the former effaces elephant difference, whereas the latter holds the animal at an impossible remove.

Between and in opposition to these dominant strands, I identify more useful resources in those images that seek to open thinking and feeling space for animal difference—drawing attention to connections, proximities, and shared histories while leaving open the gaps and uncertainties that accompany interspecies encounters. These works engender an irresolvable curiosity, a nagging and persistent sense of disconcertion, and a sympathetic but uncertain feeling for the life of other beings. Here, moving animals are kept fluid and alive. These responses are manifested in different intensities—from the gut shock of the PETA film to the enchanted sense of unease produced by Painlevé. In different ways these images provide a shock to thought, challenging cliché and affirming difference in the face of the narrow identities and affective logics that characterize mainstream representations. In focusing on elephants, I have deliberately picked an accessible and telegenic target. Nonetheless, many of the concepts and techniques that I detail can be applied to moving images of more obscure forms, processes, and interactions. The key principle remains the same: the moving imagery should open thinking spaces for a micro-politics of curiosity in which we remain unsure as to what bodies and images might yet become.

· 7 ·

BRINGING WILDLIFE TO MARKET

Flagship Species, Lively Capital, and the
Commodification of Interspecies Encounters

In the summer of 2010, the streets of Central London were full of
Asian elephants (Figure 12). Two hundred fifty fiberglass animals had
been produced for the British NGO Elephant Family, who had them
decorated by famous artists and distributed around high-profile locations. The models were accompanied by signage and willing volunteers
keen to draw attention to the plight of the species and to solicit signatures and donations. This was a spectacular fund-raising initiative.
It gained a great deal of media attention and corporate patronage
and culminated in an elephant auction organized by Christie's and
attended by royalty and A-list celebrities. The founder of Elephant
Family is Mark Shand, a charismatic adventurer, author, and brother-
in-law to Prince Charles, the future king of England. The London
Elephant Parade raised over £4 million, the profits from which to be
distributed to partner NGOs seeking to buy elephants corridors to
connect fragmented pockets of designated land in India. This effort
involves, however, the relocation and financial compensation of (often
politically and economically marginal) rural farmers.[1]

In this guise the elephant acts as an archetypal flagship species.
This is a popular political and economic strategy in contemporary
wildlife conservation, whereby a charismatic icon is mobilized to gain
funding and support. Flagships come in many guises and serve different ends, as seen with the corncrake in chapter 3. The deployment
of the Asian elephant in the Elephant Parade has all the hallmarks of
the spectacular, celebritized, neoliberal mode of conservation that has

FIGURE 12. The London Elephant Parade on the steps of the National Gallery in Trafalgar Square. Image courtesy of Elephant Family.

become prominent in the conservation industry over the past thirty years. This political economy differs markedly from what we have encountered in this book so far.

Here, the elephant has been commodified and brought to market. As aesthetic fiberglass, it becomes private and fungible. It leverages new income streams by linking markets in art and land through the liquidity of pounds, dollars, and rupees. This involves an international network of civil society organizations, elite philanthropists, scientific experts, and celebrity advocates—rather than national governments or their multilateral agencies. For advocates fiberglass elephants and other conservation commodities enable active, aspirational citizens to make a difference, incarnated as conscious and conspicuous consumers. Past market failures are addressed through the compensation of affected parties, and the elephant goes global, becoming "cosmopolitan" in its ability to link diverse and distant epistemic communities and political constituencies.[2]

In this chapter I take inspiration from the Elephant Parade to focus on the commodification of encounters between conservation publics and charismatic species. In the first section, I explore how wildlife has come to be conceived and governed as a form of "lively capital,"[3] generative of surplus value for financing conservation. I draw on growing bodies of work concerned with the commodification and financialization of the potentialities of life[4] and the creeping neoliberalization of conservation.[5] I develop these literatures by attending to the role of affect and nonhuman agency in conservation practice and explore how the specificities of the life being commodified shapes the character of lively capital. I trace how powered-up forms of nonhuman charisma come to shape what Brockington and Scholfield have termed the "conservationist mode of production."[6]

In the three subsequent sections, I explore the consequences for animals, ecologies, and environmental subjects of configuring conservation around commodified encounters with charismatic species. Here, I move from elephants to focus on the commodification of animals in captivity and the rise of "scientific ecotourism" as a flexible mechanism for funding and delivering conservation. I then return to elephants to explore the politics of flagship species conservation, focusing on situations where their cosmopolitan character is called into question. In conclusion, I review the implications of my analysis for understanding value and the politics of valuation in conservation and begin to reflect on how things might be done otherwise.

VALUING ENCOUNTERS IN CONTEMPORARY CONSERVATION

There are multiple political, economic, and ecological processes through which encounters with charismatic species might be (and have been) valued. Corncrakes, cows, and elephants can become entangled in diverse circuits of knowledge, governance, and exchange. In the accounts provided so far, I have focused on fairly esoteric and largely public wildlife encounters and associated forms of value. These have tangential relationships to capitalism and the commodity form. While these forms of value matter, they provide a rather partial and perhaps anachronistic sample of the political economy of nature conservation. Historians of the sector note significant and pervasive shifts in its dominant organizational forms and operating practices toward the end of the twentieth century, concomitant with wider changes in Western environmentalism.[7] There are three broad and interconnected trends that are especially significant for framing how encounters with charismatic species have come to be valued in prevalent modes of contemporary conservation.

The first relates to the rise of neoliberal environmentalism and associated modes of conservation. At the end of World War II, wildlife was largely understood (at least in North America and western Europe) as the concern of the State and science-led bureaucratic institutions, working closely with a set of nongovernmental organizations—on both the ecological left and the land-owning right. Conservation was serious, austere, and largely antipathetic to capitalism, the market, and the commodification of life and matter. There is a long history of commodifying animal encounters for entertainment and tourism (e.g., in circuses, zoos, cinema, hunting, and safari). It is only relatively recently that commodification has been promoted as the solution to, rather than the cause of, environmental problems. Nature would be saved in spite of or after the end of capitalism. While we can trace the persistence of this ethos in the Hebrides and at the Oostvaardersplassen, conservation changed radically (but not always consistently) in the late 1980s as part of the wider neoliberalization of environmentalism.

As biodiversity burst onto the scene as a new, global way of organizing conservation, so the State's responsibility for funding and delivery declined. In place of the State, we have seen a significant expansion (in terms of both resources and the number of organizations) in the nonprofit/nongovernmental sector. The most powerful of these have made a U-turn in their attitude toward capitalism. Working closely with public and multilateral institutions, they have encouraged the privatization of Nature, its commodification and financialization, and the creation of new markets to encourage private sector investment. They increasingly mobilize their publics less as members of civil society and more as active, ethical consumers.

The Elephant Parade offers one example of the increasingly commonplace tools through which encounters with (real and virtual) rare charismatic species are commodified for the salvation of their free-ranging kin. Other, purportedly nonconsumptive, modes of generating such lively capital include ecotourism, wildlife film, animal-adoption schemes, zoos, and the marketing of animal-related products tied to popular media. All deliver different volumes and percentages of their profits to conservation. As we shall see, there is still some ambivalence here, but selling privatized encounters to save Nature is becoming the orthodoxy in twenty-first-century conservation.

Second, commentators have noted the increased importance of popular media to neoliberal conservation. While wildlife has always been a staple of cinema and television, the volume and style of wildlife imagery have increased and shifted dramatically in recent years to better enable processes of commodification—on screen, online, and through various institutions for privatized, thematic entertainment. Virtual wildlife has flourished even as populations of living bodies decline. Critics of this trend have developed Guy Debord's Marxist critique of spectacle—or "alienated experience achieved through the fetishization of images"[8]—to explore how powerful actors in the conservation industry frame the causes, consequences, and possible solutions to conservation problems in media outputs. They argue that popular representations of conservation—like the Elephant Parade—frequently fail to articulate historical roots, unequal political economies, and consumers' complicity in perpetuating conservation problems.

Affective images of charismatic species act as a commodity fetish, encouraging a superficial and ineffectual politics insufficient for understanding and addressing the social and ecological complexities of conservation.[9] Spectacle encourages consumer-citizens to turn their backs on proximal ecologies and uncommodified wildlife encounters and get lost in commodified simulacra of nature. Anna Tsing identifies modes of neoliberal "spectacular accumulation" in which images of charismatic organisms become vital commodities for funding the, often undesirable, operations of the big environmental NGOs.[10] To date, there has been little attention in this literature to the taxonomy of these spectacles or to the specificities and machinations of the mediated lively capital they produce.

Much of this work has been concerned with spectacle in the circulation and sedentary consumption of images. We can push this further to consider the relationship between the commercialization of conservation and the rise of what Joseph Pine and James Gilmore term the "experience economy," in which businesses increasingly look beyond selling material products or useful services to orchestrate memorable events and experiences.[11] This involves significant "emotional labor" on the part of human and nonhuman employees charged with performing and orchestrating such encounters.[12] We can apply this hypothesis to make sense of a third shift in the political economy of conservation most clearly illustrated in the rise of ecotourism. Here, the value of a commodified consumer experience relates to the successful staging of authentic, fantastical, and even ironical encounters with social and ecological difference.[13] The bodies and lively performances of plants and animals are vital for their successful capitalization.

Proximal, embodied encounters with charismatic fauna are the elixir here—from the Serengeti to the zoo. Their evocative, promissory images populate marketing materials. Wildlife managers go to great lengths to ensure bountiful and visible populations (at least within the bounds of the private spaces in which they might be viewed), and their diurnal rhythms structure those of their tourist admirers. James Carrier and Donald MacLeod argue that the performance of contemporary ecotourism is similarly characterized by spectacle. Ecotourists are kept within a "bubble" comprising controlled encounters that do little to divulge and may often perpetuate the underlying drivers of biodiversity loss and social injustice.[14] Stephanie Rutherford offers a more equivocal analysis of the modes of green governmentality in North American ecotourism and wildlife entertainment.[15] She is less anxious over the purported loss of authenticity but echoes a belief in the formative power of encounters with charismatic species.

To summarize, the rise of neoliberal, spectacular, and experiential conservation, associated with the declining role of the State and the growing importance of the citizen-consumer, has increased the economic value of flagship species. While charismatic species have always been important (witness the WWF's use of the panda), they have gained greater significance as vehicles for securing funding, attracting members, granting legitimacy, and making interventions. In the three sections that follow, I reflect on some of the implications of this trend for the animals, ecologies, and human subjects governed through wildlife conservation. In so doing, I hope to arrive at a better understanding of the character, specificities, and consequences of the capitalization of wildlife in the prevalent mode of conservation emerging in and as a response to the Anthropocene.

THE AFFECTIVE ECONOMIES OF THE ZOO AND
THE CAPTIVITY PARADOX

Perhaps the most striking example of the significance afforded charismatic species in the affective economies of wildlife is the modern zoo. Zoos increasingly present themselves as vital conservation institutions—raising money, building public awareness and education, conducting research, and hosting populations of rare species.[16] Their financial success still depends, however, upon their appeal to their paying publics. The desire to educate, advocate, and conserve must be balanced with the provision of entertainment. In zoos' efforts to square this circle, a remarkably narrow and internationally convergent collection of globally rare and threatened charismatic species has come to play a vital role. Pandas, polar bears, elephants, lions, gorillas, and tigers are among the principal breadwinners in this zoological pantheon. It is rare to find an institution without some of these (or in pursuit of them), and the architecture of their captivity is configured to ensure frequent, dependable, and proximal encounters with live specimens. These animals are exemplary forms of lively capital.

The arrival of individual representatives of these species can radically alter the fortunes of a zoo. In 2011 the Royal Zoological Society of Scotland (RZSL) took a ten-year loan from the Chinese government of a pair of pandas for Edinburgh Zoo. This required a formal request by the UK prime minister and an agreement by RZSL to pay the Chinese authorities $1 million (£600,000) per year.[17] Commodified through corporate partnerships (£3,000 a year), visitor tickets (£50 for a family), an adoption scheme (£40–£500), childrens' parties (£17 for a panda picnic), and assorted panda gifts (including a £48 panda kilt), the animals have generated significant returns on this investment. In 2012 the panda accounted largely for a 50 percent (£5 million) increase in annual revenue.

Together, this enterprise saved the historically unprofitable zoo from probable extinction.[18] In order to address the inevitable drop-off in public interest (and meet their conservation objectives), the zoo is now encouraging the animals to produce cubs. Here, they are hoping to emulate the commercial success of the Berlin Zoological Garden, whose famous infant polar bear, Knut, netted the zoo some $30 million in his four-year life. During this period the zoo also turned a profit for the first time in its 167-year history. Knut is estimated to have generated some $140 million in global business and is now a registered trademark (as are Sunshine and Sweetie, the Anglicized names of the Edinburgh pandas).[19]

The economic significance of charismatic animals to zoos has implications for animal welfare. The central, tragic paradox to this commodification of charisma is that many of the animals that are so captivating to zoo-going publics are those that fare least well in conditions of captivity. This is the *captivity paradox.* Polar bears, elephants, and tigers, for example, are intelligent, sociable, wide-ranging animals poorly adapted to the confined, often solitary, and relatively sensorially impoverished conditions of captive life. They are not, as Heidegger would have us believe, truly captivated by their world. Enrolled and exploited as nonhuman emotional laborers, they get fat, lame, and bored and exhibit stereotypic behaviors—or "zoochoses."[20] As an inadvertent dividend of animal testing, they may be given tranquilizing, antidepressant, and antipsychotic drugs formally developed on laboratory kin. Many zoo animals are nocturnal or crepuscular, shy, sleepy, and habitually inclined to avoid the diurnal human contact upon which zoos depend. They must be trained or habituated to perform desired animal subjectivities through a wide range of governmental techniques, including architectural interventions and practices of positive reinforcement.[21]

Although the living conditions of zoo animals have improved significantly in recent years, it is difficult—even with a sympathetic take on zoos' claims for conservation—not to see these charismatic icons as sacrificial victims, performing their captive, commodified, and simulated lives so that other (often less charismatic, free-ranging) life might persist. In her analysis of animals caught up in and commodified through the exotic pet trade, Rosemary-Claire Collard describes captive charismatics as "undead things,"[22] lively commodities performing exotic behaviors in encounters that are fundamentally alienated from habituated social and ecological relations. In this context charisma becomes a mixed blessing. It may do enough to stall the slide to species extinction, but it imposes great suffering upon those individual animal subjects charged with perpetuating the populations and gene pools they purportedly incarnate.

LIVELY CAPITAL AND IN SITU CONSERVATION

In chapter 3, I document how flagship species have a significant effect on the scope and practice of in situ conservation in the United Kingdom despite the panoptic aspirations of biodiversity. This is true in the global north but is even more the case in the global south. In the political economy of international conservation, 90 percent of conservation funding originates and is spent in economically rich countries.[23] This money helps support the science and practical interventions described in chapters 3, 4, and 5. The remaining 10 percent of conservation funding includes the small amounts raised in poor nations and the geographically flexible resources that flow from rich to poor countries. International nongovernmental organizations have come to account for an increasing percentage of these resources over the past fifteen years, due to the changes outlined previously. During this period public spending on conservation has declined by about 50 percent.[24] In this section I focus on the character and influence of scientific ecotourism (or conservation "voluntourism"[25]) as an example of one such international funding mechanism. Here, there has been an explicit attempt to market popular, affective encounters with charismatic species as a means for financing, delivering, and governing conservation.

Advocates for voluntourism promote a win-win scenario in which active, affluent citizens in the north can make a difference to wildlife and communities in the global south by channeling forms of "global environmental citizenship."[26] This model offers northern governments an economic means for meeting their international environmental (and development) obligations.[27] It fosters environmental awareness among citizen-consumers and provides capacity and currency for conservation in countries where the state is absent or has failed.[28] There is an established and sophisticated literature in geography and anthropology that is critical of these claims, linking ecotourism to the process of neoliberal conservation and spectacle.[29]

While this work is important, it has largely neglected the importance of affective encounters with species and landscapes in animating the practice and configuring the scope of neoliberal conservation. There has been a tendency to downplay nonhuman agency in the emergence and operations of conservation, to flatten out the non-human world to a set of commodities—such that elephants and turtles or deserts and rainforests are understood as equivalent.[30]

To begin to address this gap, we can consider the results of an investigation that I conducted in 2006–7 into international conservation voluntourism from the United Kingdom.[31] This research mapped geographic and taxonomic patterns in the species and places that were the targets of voluntourists' care. Tropical forests, savannas, and coral reefs were the most popular biogeographical zones, and volunteers clustered into four regions: southern and eastern Africa; Central America and the Caribbean; the Andes and the Amazon; and Indonesian islands. South Africa is by far the single most-popular destination country, followed by Costa Rica and Indonesia.

Table 2 shows the taxonomic patterns of the species targeted by conservation volunteer programs. It demonstrates the strong preferences that exist toward mammals, especially the "big five" (the white and black rhino, African elephant, buffalo, lion, and leopard). Turtles, primates, and cetaceans also emerged as popular targets. The lion was the species subject to the most interest. There were very few voluntourism programs for plants, invertebrates, and birds. These gaps were more pronounced than in the patterns mapped in chapter 3.

A political ecological analysis of the patterns in this data and the marketing literatures deployed by the recruiting organizations helped identify the significance of colonial history (language, national parks, and research infrastructure), postcolonial trade and migration connections (visa arrangements, flights, tourist economies, the logics and dependencies of development), and the persistence of exoticized imaginations of the tropics and "Africa."[32] These factors help explain the clustering of volunteers in cheap middle-income countries—primitive enough for an adventure, poor enough to need saving, but not too risky, inaccessible, or uncomfortable to disqualify as tourism destinations. These explanations are valuable, but the biogeographical and taxonomic preferences mapped in Table 2 are more than an artifact of cultural, political, and economic drivers. The material properties of animals and landscapes and the lived and imagined experiences of encounters with them shape the patterns of neoliberal conservation. To trace their influence, I explore three affective logics that are

TABLE 2. THE TAXONOMIC DISTRIBUTION OF VOLUNTEERS ON SPECIES-SPECIFIC CONSERVATION PROGRAMS		
TAXON	TOTAL VOLUNTEERS	PROPORTION OF VOLUNTEERS (%)
Mammals	4,086	64
Big 5	1,165	18
Herbivores	658	10
Other	137	2
Carnivores	882	14
Felines	631	10
Primates	816	13
Cetaceans	428	7
Herpetiles	836	13
Turtles	763	12
Fish	307	5
Sharks	256	4
Birds	86	1
Invertebrates	26	0.4
Caterpillars/butterflies	19	0.3
Plants	13	0.2
General rehabilitation	988	15

performed within the commodified interspecies encounters that characterize voluntourism. They are populist concerns for spectacle, touch, and adventure. These are not exclusive but emerged as the most significant in this study. They intersect with the logics mapped in the previous chapter and can be compared with the more esoteric logics of epiphany and jouissance outlined in chapter 2. These logics shape the sensuous interspecies relations through which wildlife encounters come to be valued and capitalized.

More than one-third of voluntourists covered in the survey worked on species-specific programs that focused on animals that afforded the possibility of spectacular encounters in spatially extensive landscapes—primarily, the big five in southern and eastern African savanna or cetaceans in various oceans. Raised on a visual diet of natural history documentaries, volunteers explained their desire for visual encounters with abundant populations of large animals living in uninhabited spaces. The "fingery-eyes" and awe-some logic of wildlife documentaries made wilderness paramount and human absence the currency of authenticity. Killing, eating, sex, and other social behaviors were in high demand.

This mode of engagement has its roots in hunting and safari and involves tracking, identifying, and witnessing.[33] This is an embodied vision involving disciplined scanning eyes and assisted by visualization technologies like binoculars, cameras, radio trackers, GPS, and moving vehicles. Observing and participating with volunteers spotting and tracking elephants from jeeps in a national park in Sri Lanka, it was clear that volunteers sought the sublime—privileged, private encounters with exotic alterity to be captured in photographic trophies. To differing degrees, this led to wonder and curiosity as they became more involved in the research for which they had volunteered.

In a second and somewhat different affective logic for conservation, voluntourism values touch and haptic encounters with other animals. The desire here is for proximal encounters with individual animals outside the menagerie of familiar domesticates. It involves personalized relationships that might be built up through frequent corporeal interaction—feeding, stroking, washing, playing, conversing, or just observing. This logic helps account for the popularity of large mammal rehabilitation and captive breeding projects as well as those that permit touch in the wild—for example, turtle and meerkat conservation projects.

Volunteers who traveled to Sri Lanka to work with captive elephants celebrated the chance to work in close proximity with a visually familiar but corporeally and ethologically strange animal.[34] Touching encounters allowed the direct delivery of care to a named individual elephant and offered the possibility of achieving reciprocal recognition and communication. Haptic processes of feeding, cleaning, stroking, and riding their elephants were afforded great importance during their lengthy and repeated encounters. Spectacle and vision mattered much less than intimate, immersive encounters in the flesh. Legitimate questions could be asked about the efficacy of many of the rehabilitation and captive breeding programs supported by these volunteers, which seem as complicit in the captivity paradox as zoos. Few animals ever return or become free-ranging populations.

If savanna habitats and their animals afford extensive spaces for wildlife spectacle, then tropical forests are mobilized by voluntourism operators and their clients as arenas for adventure. In visiting tropical forests, voluntourists are encouraged to "get out of their comfort zone" through physically challenging tasks, encounters with the unknown, and the privations of field life. The target organism matters less than the alien and excessive nature of the habitat. The affective logic at work involves a mixture of both adrenaline (hacking through the jungle, avoiding snakes) and quasi-puritanical self-denial (roughing it as catharsis). These places are understood as both exotic and sublime— affording heat, wildness, and space for self-realization—and as pathological spaces of disease, abjection, and moral degeneration. Here, voluntourism bleeds into a mode of "dark tourism" that explicitly plays on the risks associated with its destination and activities.[35]

Together, these three affective logics help shape the scope of conservation delivered through market-oriented voluntourism. Biologists involved in delivering conservation programs dependent on volunteers spoke of the challenges of catering to these desires. Much of the fieldwork and analysis upon which good conservation science depends requires routine, repetitive, and databased activities conducive to intellectual satisfaction but not adventure, touch, or spectacle. In science an absence can be as significant as a presence. But it is difficult to convince a fee-paying volunteer of the merits of an absence when that absence is the elephant they are desperate to encounter. Many volunteers find field science boring and complain if compelled to spend time at a computer. As with the choreography of encounters in captivity, biologists find ways to reconcile research, conservation, and entertainment. They might divide their year into periods for entertaining volunteers and periods without volunteers for doing research. Or they work hard to narrate, value, and dramatize the conduct of science while ensuring volunteers are compensated with premium encounters with wildlife in protected areas beyond those available to normal ecotourists.

Many voluntourists do travel in bubbles, consuming simulations, largely unaware of the histories they are performing and perpetuating through their encounters. But many do so sincerely. It is too cynical to dismiss them outright on these grounds. It also offers a limited analysis of what is taking place in neoliberal conservation. These tourists paid, traveled, and endured because they cared. The road to hell may be paved with good intentions, but attending to the affective logics in these interspecies encounters helps explain some of the emergent patterns and the power of encounter value in neoliberal conservation. The care expressed by voluntourists matters and can have fraught and promising material consequences.

LIVELY CAPITAL AND THE POLITICS OF FLAGSHIP SPECIES CONSERVATION

Max Weber understands human charisma as a "gift of grace": a valuable yet threatened property that helps achieve organizational order in an increasingly "disenchanted" and bureaucratic modern world.[36] Charismatic flagship species have similar effects. They provide an enchanting, unifying catalyst for organizational activity. Successful flagship species should act as lively "boundary objects" that unite diverse political and epistemic communities in pursuit of a common program.[37] This was the case with the corncrake in chapter 3, where crofters and conservationists were able to reconcile their political, economic, and epistemological differences to secure the future of the bird and the agricultural system upon which it depended. Here, the corncrake became an icon for Hebridean machair conservation.

Geographers and anthropologists interested in more spatially extensive, culturally diverse, and politically fraught networks of environmental governance have critically examined the comparable concept of the "cosmopolitan species."[38] Here, a cosmopolitan/global flagship would be understood the same the world over. Unlike a boundary object, which seeks to reconcile political and epistemic difference, a cosmopolitan represents a universal norm—a global desire for biodiversity conservation and the focal species' salvation. In this section I deploy the concepts of charisma, boundary objects, and nonhuman cosmopolitans to interrogate the politics associated with the commodification of encounters with flagship species. To do so, I return to elephants and the international networks performed through the circulating fiberglass icons of the Elephant Parade. In contrast to the consensual politics of corncrake conservation, I want to examine situations of epistemic and political difference where the value of some encounters with elephants has a negative valency.

In his extensive ethnographic investigations of human–elephant conflict in Assam, Maan Barua has traced what happened when the monies gathered in the London Elephant Parade touched down in India.[39] As their fiberglass kin settled into their new lives in the galleries, houses, and shopping arcades of London's rich and famous, so Elephant Family transferred funds to their partner organization—the Wildlife Trust of India. They began to implement a new round of land acquisition that formed part of Project Elephant—a larger, long-planned, and part state-funded project to secure "elephant corridors" to protect the national population.[40] These would connect fragmented pockets of legally protected land to enable the movement of elephants and ameliorate the damage done to the agricultural areas through which the animals frequently passed and lingered. The aim is to buy land from the villagers (compulsorily if need be), create compensation schemes for agricultural damage, and improve fencing and other elephant-deterrence technologies.

So far the elephant remains cosmopolitan; it returns to India as a species of global conservation concern in need of immediate and drastic action. It experiences unfettered international mobility and popular support by virtue of the strongly aligned values of the elite, postcolonial wildlife enthusiasts and urban publics in London and Delhi who fund and run these NGOs. The charisma of elephants is valued differently, however, by many of the farmers living in or alongside the proposed elephant corridors. In Assam, Barua focuses on encounters between farmers and villagers and a small herd of "rogue" male elephants who have become accomplished crop raiders—venturing into villages, often at night, in pursuit of paddy (rice) and illicit alcohol. He documents how such encounters with elephants, which are increasingly common as agriculture expands, frequently result in significant damage to farmers' livelihoods, property, and mental and physical welfare. Barua explains that "in India alone, elephants kill c. 400 people every year, damage 10,000–15,000 houses and destroy crops worth up to US$3 million annually,"[41] and the number of human deaths in Assam due to conflict with elephants has doubled in the past decade. People live in fear, deprived of sleep by the need to guard their crops from nocturnal forays and anxious over their precarious livelihoods. The value of an interspecies encounter has a negative valence here in comparison to the enthusiastic metropolitan consumers of the Elephant Parade or the ecotourists whom the national parks and corridors hope to attract.

On the ground in Assam, the elephant ceases to act as a boundary object, and the unequal politics of its incarnation as a commodified global cosmopolitan becomes manifest. As with Annu Jalais's reflections on tiger conservation in the Sundarbans, the threatening of abundance, not the distant global threat of extinction, is the human experience.[42] A small number of politically and economically marginal citizens are being tasked with bearing a deadly global burden. Powered up and rendered fungible as affective lively capital, the global elephant effaces such local specificities. As money it flows like mercury to secure cheap and distant land.[43]

Assamese villagers have resisted schemes to increase the depredations of pest species. They have disputed the acquisition of land for the corridor through various means of nonviolent protest and nonparticipation. Barua explains how the elephant becomes understood as a government-tolerated *dacoit,* or "bandit," a violent operative serving to clear people from land that will be annexed to serve elite interests in charismatic wildlife. This is a familiar critique of conservation in South Asia. In a further political strategy that reworks the logics of the cosmopolitan elephant, Barua reports how villagers have taken to naming rogue elephants Bin Laden—even tagging one animal that had been killed while crop raiding. Media-savvy opponents of the expansion of elephant territory invoke the threat of this charismatic global-terrorist villain to draw support for their campaign.[44]

ENCOUNTER VALUE

In this chapter I develop my analysis of the shifting character and political and economic significance of nonhuman charisma in contemporary conservation. Building on the discussions of chapters 2 and 3, I trace relations through which encounters with charismatic species are commodified to generate forms of lively capital to fund conservation in an era of declining public sector support. In contrast to most of the forms of lively capital that have been mapped in existing critical scholarship, here valued life takes the form of individual organisms or their aggregations into populations of species. Somewhat paradoxically, their valued and commodified liveliness relates less to their generative potential to deliver "services" and more to their aesthetic power to enchant through "authentic" encounters.

Mapping the specificities of this form of lively capital helps to flesh out the characteristics of value in certain currents of contemporary conservation. I have focused on encounters, and it is therefore appropriate to bring this discussion into conversation with Donna Haraway's writings on "encounter value." In chapter 2 of *When Species Meet,* Haraway examines the operations of lively capital in the North American dog-keeping industry. Through her analysis of dogs as commodities, she identifies the importance of encounter value, which she presents as a supplement to relational Marxist conceptions of use and exchange value. Encounter value describes the value that accrues from multispecies encounters, recognizing the agency (even perhaps the labor) of other life-forms. Such encounters are not always useful, nor are they accurately (or for Haraway desirably) valued through metrics of universal exchange. She suggests that multispecies encounters have values whose qualities and ethics are poorly served by the formation of commodities, and thus a third type of value is required.

Through the analysis performed in this chapter, we can connect the concept of encounter value to that of nonhuman charisma and its commodification through the generation of exchange value. Encounter value—like charisma—is a relational concept. It relates both to the properties of the organism being encountered and the affective logic that shapes any encounter and how it is valued. It therefore comes in many forms. For example, encounters with Asian elephants are valued very differently by scientists, voluntourists, mahouts, and Assamese farmers. Few find the elephant instrumentally useful, but all grant the elephant different forms of intrinsic value as a living being. This plurality of encounter values is challenged by the introduction of money as a universal medium of exchange between all four of these elephant interest groups. Once encounters have exchange value, their specificities are effaced, and they can be "objectively" compared, ranked, and governed accordingly, such is the logic of cost-benefit analysis. In aggregate terms, under the popular and commercial affective logics of spectacle, touch, and adventure that dominate contemporary conservation, it appears that certain encounters with Asian elephants are more highly valued than others. Meanwhile, encounters with elephants are more valuable than those with corncrakes, and encounters with both are afforded greater value than those with the vast majority of plant and invertebrate species.

This recognition of the qualitative differences between encounter value and use and exchange value in conservation enables critical evaluation of the rise of neoliberal conservation. Valuing encounters beyond their instrumental services suggests that we need to examine the interspecies ethics of conservation in the context of the biopolitical power of conservation in action. As this chapter demonstrates, charismatic species shape the scope and conduct of conservation—both in situ and ex situ—with important material consequences for the people, animals, and wider ecologies subject to the forms of governance they inform. Popular forms of charisma have been powered up by the arrival of market conservation and the generation of novel forms of lively capital.

If, as current trends suggest, conservation maintains and extends its dependence on market mechanisms, then we can anticipate some important implications of this analysis. The natures that lively capital can see[45] (and feel) in the form of commodified encounter value face an ambiguous future. On the one hand, the demand for captive encounters with charismatics should see their future perpetuation as genes and species in zoos, even as the globalizing black market in illegal animals that exists to meet the demand for private pets empties out the forests and other habitats in which they reside.[46] On the other hand, the future life for the captive "undead things" who embody these breeding populations looks rather bleak, destined by their captivating nature to a life of debilitating captivity. If we take these animals as nonhuman emotional laborers, we might even understand their captivity in Marxist terms as a mode of exploitation.

Less captive but equally captivating charismatic species might fare better than their noncharismatic, undomesticated, or nonresilient kin. Markets exist now to support extensive areas for conserving and breeding some charismatic wildlife. Landscapes subject to a conservationist mode of production are being configured around the delivery of such commodified encounters. The demand for such encounters has a generative logic and is materially transforming the reproductive ecologies of wild spaces. For example, the wildlife in some of the big-game reserves of southern Africa has been subsumed to the logics of spectacle that drive hunting and safari. As with the public practices of conservation reported in chapter 3, there is a virtualism at work here in which ecologies are emerging with heightened degrees of nonhuman charisma.[47]

Uncharismatic species continue to suffer. The logics of this market generate an even more tightly specified set of desired targets, shutting out more obscure species and esoteric relations. Many of the most charismatic species remain in decline. There is a further paradox here that many of those animals that emerge as cosmopolitan flagships seem to be those with which modern agricultural, urban, and industrial systems live least well—tigers, elephants, polar bears, and pandas, for example. Their futures cannot be ensured through the commodification of encounters to pay for protected areas. A more systematic shift is required.

In her original writings about lively capital, Haraway is hopeful that an attention to encounter value might offer better forms of companionship. One approach that is familiar in critical work on commodification in political ecology would be to seek out unalienated (i.e., uncommodified) encounters and to celebrate ways of valuing these encounters that need not make recourse to the logics of market exchange. To a degree this approach is implicit in the contrasting tones in which I discuss my research subjects in this chapter and in chapters 3, 4, and 5. On reflection I have found more sympathy with the scientists and crofters in the Hebrides and the polders than with the international voluntourists and market conservationists. But an explicit and totalizing disavowal of commodified encounters seems excessive. It both smacks of a residual romanticism that elevates the premodern as authentic and writes off some of the sincere and occasionally life-changing encounters made possible by commercial television, tourism, and other forms of wildlife entertainment, as discussed in the previous chapter. As seen with crop-raiding elephants, not all authentic encounters are desirable. Nor are all commodified encounters flawed.

The recognition of encounter value flags the human politics performed by its commodification, attending to whose values are performed in situations of discord. In the examples briefly discussed here, we find much more unequal and undemocratic relations than those we encountered in the Hebrides and the Netherlands. While there was discord in the Netherlands, it involved disagreements within a national citizenry with recourse to legal and political processes about matters with limited impact on their personal or economic security. The global networks that link the commodified Asian elephant with Assamese farmers and the British aristocracy are democratically opaque and economically unequal. This unsavory politics is energized and exacerbated by market environmentalism. Attending to encounters helps unmask these operations, but alternative modes of valuation are required.

SPACES FOR WILDLIFE

Alternative Topologies for Life in Novel Ecosystems

I took the Figure 13 photograph fifty stories up, on top of an office block in Canary Wharf in Central London. Alongside me was Dusty Gedge, an urban ecologist and cofounder of Livingroofs, an organization dedicated to "greening" UK roof spaces.[1] Around us higher still towered the headquarters of corporate behemoths in the global financial services industry. This is a landscape of sheer edges and abysses, of concrete, glass, and steel. It looks and feels a long way from Nature—or even the natures that have so far been featured in this book. But Dusty was grinning, for underfoot there was a scrubby carpet of bare soil and brownish vegetation. When he crouched down to examine a pitfall trap, it was teeming with beetles and myriad other invertebrates. He explained how ballooning spiders blown across the Channel and funneled to the city were finding habitat for British colonization. Incongruous as it seems, this site is a verdant oasis for wildlife and adaptation, set right among some of the most densely populated and expensive real estate in the country.

Later that summer, I was in a neglected graveyard in South London. It was dusk, and I had joined Helen, another urban ecologist and entomologist, on a tour of the city's deadwood ecologies. Rush hour traffic thundered past on a nearby trunk road. The rotting fallen trees were ornamented with a flotsam of crisp packets and drink cans. The air was dank, and we were eyed suspiciously by a couple of homeless men. Nonetheless, Helen was overjoyed. The graveyard was flush with invertebrate life, and she spotted stag beetles, *Lucanus cervus* (Figure 14), swarming over a dead log. These large, rare beetles were

FIGURE 13. A living roof on top of a fifty-story building at Canary Wharf in Central London. Photograph by the author.

emerging from their three-year larval stage. Cumbersome and alien, they launched themselves into the gloom. For a few short weeks, they would fly, fight, and breed before searching for further fragments of urban decay in which to lay their eggs.

Helen and Dusty are part of a loose alliance of unorthodox urban ecologists, conservationists, and politicians. During the past thirty years, they have been exploring and celebrating the United Kingdom's "unofficial countryside."[2] These urban wilds range beyond much-loved parks and gardens to encompass the "edgelands" of abandoned brownfields, railways, canals, and other oft-neglected fragments.[3] Their work has flagged the ecological abundance, originality, and significance of these ecologies. It promotes their local climatological and hydrological importance. It even claims their agricultural potential. The cultural and political value of urban ecologies in a majority-urban world is

FIGURE 14. A UK stag beetle. Image reproduced courtesy of William Harvey,
http://william harveyphotography.co.uk.

identified, with proximal places promoted for "connecting" urban citizens with "nature" and improving mental and physical health. This
politics interfaces at times with long-standing arguments for environmental justice.[4] These diverse political ecological enthusiasms do not
always align, but they have put urban wilds in vogue.

Here, I want to suggest that these enthusiasms can be understood
as part of a wider shift that is taking place in the geographies of nature
conservation. A reappraisal of urban ecologies is part of a growing
interest in nature outside protected areas. New buzzwords in conservation include *green infrastructure, ecological corridors,* and *networks.*
These will permit landscape-scale "connectivity," "permeability," and
"fluidity." Reserves will be connected; organisms, translocated; habitats, created; and wider landscapes, managed to enable adaptation to
a changing climate. In short, conservationists are finding that wildlife
is poorly understood and governed by drawing and policing boxes.
Wildlife is not out there, mapped to and fixed within the wilderness
or the countryside. Instead, wildlife is in here—in cities, in gardens,
and even in our bodies. It is also on the move—transgressing national,
regional, and other territorial boundaries, performing diverse and

discordant animal, plant, and other nonhuman geographies. These movements are both desired and feared (for example, in anxieties over the biosecurity risks of ecological and pathological invasion).

In this chapter I focus on urban conservation in the United Kingdom in order to provide an explicit analysis of the spatial dimensions of wildlife and its conservation. In chapter 1, I argue that an ontology of wildlife necessitates a relational, topological approach to space. I suggest that conceiving of wildlife topologically helps attune to the multiple spatialities associated with the more-than-human geographies of conservation. Topology helps in understanding the mobilities of wildlife beyond the Cartesian cartography of the familiar topographic map. Here, I explore how a topological approach helps identify the territorial trap into which the orthodox biogeographies of conservation have fallen. It also enables a critical examination of novel spatial relations associated with the rise of urban and landscape-scale conservation. To explore these, I draw on and develop alternative topological metaphors for guiding wildlife conservation after the Anthropocene.

NATURE CONSERVATION AND THE TERRITORIAL TRAP

Since its inception, nature conservation has sought a territorial solution to the loss of biological diversity. Great energies and resources have been invested in securing, policing, and legitimating a portfolio of protected areas. There is an extensive and growing conservation estate, now comprising at least 10 percent of the earth's land area.[5] This is an impressive political achievement that has helped avert significant decreases in the diversity and abundance of wildlife.[6] But these territories tend to adhere to a restrictive geography. Protected areas are generally conceived as discrete and bounded spaces. They are seen as isolated within and threatened by an inhospitable surrounding landscape matrix comprising urban, industrial, and agricultural land use.

This spatial imagination has diverse and entangled genealogies, including aristocratic and colonial hunting reserves, romantic preoccupations with sublime landscapes designated as national parks, and religious prohibitions associated with the protection of sacred groves.[7] These territorial practices were energized in the twentieth century by the influence of the theory of island biogeography on conservation. Early architects of what became known as conservation biology—like E. O. Wilson and Robert MacArthur—focused their research on undisturbed islands.[8] Such islands became laboratories for exploring species–area relationships. From these island laboratories emerged the prevalent conceptual toolkit for conservation. Conservationists applied this theory to a range of protected areas, which tended to be visualized as islands adrift within a sea bereft of ecological import, especially for those concerned with terrestrial plants and animals.[9]

The North American, European colonial, and (to a lesser extent) continental European cultural histories of nature reservation endowed late twentieth-century conservationists with an estate concentrated on territories marked by the absence of modern humans. These became the islands for conservation management. As a range of critics (including both social scientists and ecologists) have argued, these biogeographies for the conservation of Nature tend to purify space and stabilize time. Protected areas map the modern Nature–Society binary to establish and police fixed and ranked territories for Nature. They inform violent and iniquitous practices of "fortress conservation"[10] in which marginal people are often evicted and subsequently excluded from common land. They frequently exoticize distant and inaccessible natures and relegate proximal, mundane, and more accessible ecologies.[11]

Here, the wilderness, the nature reserve, and fragments of the premodern countryside are elevated. Such sites are set against (sub)urban areas, which are relegated to "ecological sacrifice zones."[12] Otherwise divergent romantic and Enlightenment conceptions of the city agree on the separation of the urban from the wilderness/countryside. The former is elevated or disdained as the site of industry, flux, and civilization, whereas the latter is reviled or desired as a site for wildness, temptation, and tradition.[13] As a consequence, Steve Hinchliffe and his coauthors argue, urban ecologies have been long neglected by nature conservationists: "Not pure enough to be true and not human enough to be political, urban wilds have no constituency."[14] Wildlife is out of place and out of sync with the city, disregarded or disdained as domestic, feral, or risky.

This static, frequently binary biogeography is founded upon a "regional topology." Anne-Marie Mol and John Law explain how this entails

> a version of the social [and by extension the natural] in which
> space is exclusive. Neat divisions, no overlap. Here or there,
> each place is located on one side of the boundary. It is thus
> that an inside and an outside are created. What is similar is
> close. What is different is elsewhere.[15]

Understood this way, conservation is caught in a "territorial trap"[16] in which nations, nature reserves, and other politicalized units become the bounded containers for Nature. Wildlife transgressing or living outside their boundaries is deemed lost. Management has tended toward fixity, preempting and forestalling ecological processes in the interests of preservation and/or biosecurity. Evidence of this topology is in the biopolitics of corncrake conservation in the Hebrides, as seen in chapter 4. Here, conservationists seek to secure the future of the species and the premodern practices of crofting by fixing the ecologies of their island laboratories.

URBAN WILDLIFE

A static regional topology is of limited use to urban conservation. Urban ecologies are archetypal novel ecosystems. They have emerged recently in relation to the deep time of plate tectonics, geomorphology, and allopatric speciation, as well as the (largely nonurban) history of human civilization. Large, high-rise, and networked cities with dense concentrations of human inhabitants contain unusual and often unprecedented ecological, climatological, and geomorphological forms and processes. Cities have high concentrations of resources and levels of productivity. They have distinct and generally accelerated disturbance regimes. They are far from equilibrium.[17] Urban ecologies are well connected and have high levels of human and nonhuman mobility. They comprise cosmopolitan species assemblages, including a set of "synurbic species." These are organisms that have become urban, flourishing within or adapting to urban ecologies.[18] Geologists have argued that urbanization generates novel strata. Urban ecologies will persist in the fossil record and have been discussed as one of the signature features (or "golden spikes") for the diagnosis of the Anthropocene.[19]

Cities therefore offer urban ecologists laboratories to study in extreme form the social, ecological, and biogeographical changes associated with globalization in the Anthropocene. For conservationists, however, cities matter because they are where the majority of the world's population lives. They are centers of economic, political, and cultural power, key nodes in networks of global governance. They are also the sites of "everyday environmentalism" in which many people interact with the nonhuman world.[20] The cosmopolitan, mundane, and vernacular of cities reflect the diversity of their human populace. They bear traces of histories of trade, immigration, cultivation, escape, and abandonment. In parks, gardens, and assorted wastelands, people connect with a more-than-human ecology. There is hope that these encounters can be recognized and channeled to catalyze new forms of environmental responsibility and citizenship. For advocates like those I describe in this chapter, urban wilds are promoted as archetypal spaces for conservation after the Anthropocene.

Although there have been significant recent achievements, this is still a hard sell. To make a case for wildlife in the city, urban ecologists like Dusty and Helen have had to counter the prevalent regional topology. They have also had to deliberate with competing forms of "green urbanism."[21] This has involved a number of spatial strategies. The first has been to reverse the terms of the binary opposition and flag the higher levels of biodiversity to be found in some urban sites relative to the surrounding intensive arable countryside. For example, to avert the redevelopment of an ecologically rich postindustrial site at Canvey Island on the Thames Estuary, entomologists and urban ecologists at Buglife (see chapter 3) promoted the site as "Britain's brownfield rainforest."[22] They surveyed and publicized an abundance of rare insects and plants, which was favorably compared to their absence in the surrounding "agricultural desert." This counterintuitive story caught the public attention, and the campaign was successful. In 2005 the site became the first brownfield in the United Kingdom to be legally protected specifically for its invertebrates. It is now a nature reserve.

But this is a rare success story. Neighboring brownfields along the Thames Estuary (now marketed as the Thames Gateway) have continued to be tidied up or put under concrete. In the late 1990s this area was earmarked for development under the New Labour government's Urban Renaissance. This regeneration program targeted brownfield land to build compact, "green," and "sustainable" cities and prevent expansion into the greenbelt.[23] Campaigners in London explained how the temporalities and aesthetics of many of their valued urban ecologies did not conform to this green urbanism. Many of these ecologies—like Dusty's living roofs or Helen's deadwoods—are ruderal or saproxylic. Ruderal (from the Latin *rudus*, meaning "rubble") describes the species that are first to colonize disturbed/abandoned land. Ruderal ecologies are common on postindustrial land and are reliant upon its regular disturbance. Saproxylic ecologies and their valued fungi and invertebrates are dependent on decaying wood. They require patience or neglect.

In an urban context such sites are generally figured as unaesthetic wastelands. As one London-based urban conservationist put it to me:

> Brownfield sites look ugly. Though quite often they are very flowery, they don't fit the image of the rural idyll with hedgerows and rolling hills and woods and streams and lakes. There aren't any Constable pictures of derelict rubble-strewn landscapes. Brown is the color of dirt, and they are seen as dirty and grimy and fly tipped with rubble. Not very nice.

Here, the static, regional topology of the ruralist English landscape imposes a ranked and color-coded temporality on wildlife. Purportedly timeless green spaces are valued at the expense of ruderal brownfields and rotting wood, whose lively inhabitants are deemed to be out of place and out of sync.[24]

The dynamic ecologies of ruderal sites are often seen as both risky and unproductive. Their ecologies of abandonment are associated with social groups and practices neglected by urban capitalism. They are cast as feral spaces for illicit or devious activities in need of social control.[25] As a consequence many brownfield sites that escape redevelopment tend to be "restored." According to the logics of green urbanism, this requires importing topsoil and planting grasses and trees to create open parkland landscapes. These have greater amenity value and ensure easy visibility for public surveillance.

For Annie Chipchase and Matthew Frith, urban ecologists at the London Wildlife Trust, this amounts to "greenwashing."[26] Here, ruderal and (to a lesser extent) saproxylic ecologies are sanitized, effacing their ecological value so that they conform to a ruralist ideal. Commentators on trends in urban gardening note similar trends in the rationalization of domestic natures—including the popularity of decking, lawns (or even AstroTurf), and amenable plant species.[27] Ironically, this greenwashing mentality can do as much damage to wildlife difference as wishing to make brownfields gray.[28]

Deadwood creatures like the stag beetle are anathema to the modern hygienic city, whose defining features are accelerated change and the control and banishment of unwanted nonhuman entities and processes. Such saproxylic insects require rot and abandonment. Their habitats serve as a constant reminder of mortality and the inevitability of entropic decay. Unlike synanthropic species like bedbugs, termites, pigeons, and rats, which flourish within the mobilities of contemporary urban life, stag beetles require slowness and consistency to complete their life cycle. They are out of sync with the creative destruction of the city.

To counter the persistence of a regional topology and its encroachment into the unofficial spaces for wildlife in the city, urban ecologists have sought to develop alternative ways of conceiving and governing the geographies of wildlife that are analogous to a fluid topology. Mol and Law differentiate a fluid from a regional topology as follows:

> There is a sameness, a shape constancy, which does not depend on any particular defining feature or relationship, but rather on the existence of many instances, which overlap with one another partially. So there are no great breaks or disruptions. Instead there is a process of gradual adaptation. Shape invariance is secured in a fluid topology in a process of more or less gentle flow. It is secured by displacement which holds enough constant for long enough, which resists rupture. A topology of fluidity resonates with a world in which shape continuity precisely demands gradual change: a world in

which invariance is likely to lead to rupture, difference, and distance. In which the attempt to hold relations constant is likely to erode continuity. To lead to death.[29]

A fluid topology seeks to grant more lassitude to change and the dynamics of the relations that come to define the identity and location of an object over time and space. It thus gives scope to the difference of novel urban ecosystems and the nonlinear nonhuman becomings that I associate with an ontology of wildlife. It allows for hybrid ecologies inhabited and shaped by urban citizens and sensed as vernacular natures.

We can find expression of this fluid, immanent understanding of urban ecological value in the design and management of urban brownfields and their aerial surrogates in "extensive" living roofs.[30] Here, conservation is open-ended. Attention is directed largely at the soil to create diversity in the basic substrate. Rather than seeking to fix the roof at a moment of equilibrium by forcing it toward a final assemblage of plants, these ecological complexes are allowed to develop through colonization by local flora and fauna.[31]

In part this fluidity relates to the limited resources available to urban conservationists and the inaccessible and often invisible nature of the roofs themselves, which prevents protracted habitat management and aesthetic oversight. It also suggests, however, a deeper difference in ecological management philosophy. As Dusty put it:

> Brownfield species are opportunistic, and I am quite an opportunistic person. You create opportunity, and nature fills it, especially in an urban situation where things are always changing. That is the changing dynamic of the city, and that is how urban conservation has to work. It is ironical that nature is always changing but most of conservation is about holding fast; it is conservative.

Dusty explains that the chaotic nature of urban process and ecology makes it virtually impossible to predict a static end state for a brownfield site. Instead, he advocates for extensive roof spaces with room for the expression of biological processes.[32] There are similarities with the process-orientated ethos guiding the desired management strategy for the Oostvaardersplassen (OVP). But in this case urban ecologists are much less constrained by the cultural benchmarks that have come to configure the legal infrastructure for European conservation designation, as encountered in chapters 4 and 5. Made, feral urban sites give much more scope for experimentation than do rural wilds.

The coherence of valued urban ecologies rests, however, on managing rates and tendencies of change. The fluidity encouraged by brownfield conservationists is never totally open-ended or equally negotiated. Living roofs often require some basic management to produce their broadly defined desired ecological outcomes. For example, "invasive" species like buddleia are judged to be overly competent in their colonizing ability, to the detriment of the greater diversity of the roof itself. Trajectories toward homogeneity are averted through regular weeding. In the mainstream discourse of invasion biology, such management would be framed in the discourse of biosecurity, where the accommodating ecology of ruderal sites for nonnative invasion presents weak spots in the territorial integrity of a national flora and fauna. The urban ecologists to whom I spoke were often unconvinced, however, by the fixities imagined by such nationalized territorializations.

Meanwhile, the vitality of larger brownfield sites—like Canvey Island—often requires regular disturbance to maintain the diversity of niches for plants and insects. In the absence of large (nonhuman) animals, as at OVP, the instigators of these disturbances are usually human. On private land they are also often illegal. For example, at one site on the Thames Estuary, local ecologists had reached a tacit agreement with local track bike–riding youth. Tolerating trespass to ensure disturbance was saving fortunes in ecological labor—the costs associated with maintaining early successional habitats in rural nature reserves like the Hebrides.

The fluidity of these urban ecologies is both refreshing and risky. Dusty's opportunism extends to making brownfields fungible, permitting destruction and replacement as mitigation within the planning system. Such thinking is very much in keeping with the emerging market logics of biodiversity offsetting that are currently being embraced in the United Kingdom.[33] Market fungibility permits frequent and accelerated change and spatial dislocation. It might threaten the continuity and gradual adaptation associated with a fluid topology. As a counterpoint to fluidity, Law and Mol identify a further topology of fire. They explain:

> As with fluid constancy, movement rather than stasis is crucial. Without movement there is no consistency. The difference is that, whereas in fluidity constancy depends on gradual change, in a topology of fire constancy is produced in abrupt and discontinuous movements.[34]

In the context of ecology, we might understand fire as the topology of creative destruction, where persistent catastrophes are required to ensure continuity.[35] But a topology of fire could also describe the dramatic change that is commonly associated with the capitalist city. In a resilient, connected ecology, fire is a vital process. It is less clear that valued but fragmented urban ecologies—like those that support the stag beetle—can tolerate such consistent change. Managing the discordant multispecies harmonies of urban ecologies involves channeling flows, while playing with fire.

CONNECTIVITY

Concerns about resilience and the ecological risks of fragmentation are informing a wider reappraisal of the regional topology of protected areas among conservation biologists. As a consequence of some of the trends identified in chapter 1, there is a growing appreciation that fluid, processural, and hybrid approaches to wildlife require new ways of thinking about spatial relations. *Connectivity* is a new watchword for conservation.[36] Ecologists have reevaluated the significance of the "matrix" around protected areas both in terms of its residual ecological value and its roles in assisting species movement,[37] while the growing awareness of the biogeographical implications of accelerated climatic change has flagged the need to imagine more dynamic and porous conservation territories.[38] There is a growing sense that the current conservation estate could fast become anachronistic.

The principal metaphor used to describe these emerging topologies of landscape-scale connectivity is the network, alongside variations on this node-and-link-based spatial imagination. A biogeography of discrete territories is being overlaid with maps of intersecting lines of nonhuman mobility. Conservation planning talks frequently of ecological networks, green infrastructure, ecological corridors, and stepping-stones. These will connect fragments of valued habitat and conserve ecologies of continuous mobility. Linear landscape features like rivers, canals, railways, and even roads are being reimagined as important vectors for or barriers to nonhuman movement and navigation. As explained in chapter 1, there is a contested enthusiasm for species "translocation" and "assisted migration"—performing lines of movement in an otherwise impermeable landscape to enable adaptation to anticipated "climate envelopes." There is a growing sense that natures need to be on the move.

Network topologies have been the subject of much inquiry in geography and other areas of the social sciences. They are actively promoted by those exploring the geographies of globalization and forms of governance, trade, and nonhuman movement that cut across familiar territories.[39] We see such networks in operation in the previous chapter. This work is often informed by an understanding of networks drawn from actor–network theory, which offers a topology of lines and nodes or connections between fairly stable entities. Subsequent critics have argued that this topology is too fixed and preoccupied with bounded mobile forms, rather than the dynamic processes that bring them into being. For example, in place of the network, Tim Ingold has proposed the metaphor of the meshwork, a fluid but still reticular topology of entangled lines and knots—rather than nodes and connections.[40] His meshwork is akin to Deleuze and Guattari's figure of rhizome.[41] We can draw on these different reticular topologies for analyzing connectivity in conservation.

Both Dusty and Helen visualize their urban conservation activities within a wider landscape. For Dusty the emergence of ruderal brownfield ecologies on living roofs requires their connection to sites elsewhere. Linear, terrestrial corridors are difficult to engineer in the vertiginous landscapes of the city, so roofs are designed to accommodate and encourage colonization by other vectors. These include flight, wind, and deliberate and inadvertent human introduction. For example, it is hoped that the nonnative ballooning spiders I describe at the start of this chapter might make inner-city London a launching pad for wider British colonization—billowing out to other roofs and abandoned sites and into the wider landscape. Here, the roof becomes a welcome node for adaptive, self-propelled, and benign immigrants: no visa required for such entrepreneurial subjects. A meshwork of living roofs and other spaces allows for differentiation and international mobility. It enables free passage for novel ecologies between the city and countryside, transgressing spatial binaries.

Dusty also notes, however, that the fluidity inherent within this reticular topology is threatened by the greenwashing tendencies described earlier. A discourse of green infrastructure can serve to enact the same color-coded temporality within a networked topology. For example, developers often insist that visible living roofs are made green to conform to a gardenesque aesthetic. Planted with grass and shrubs, they are rendered stable. Scaled up across the city, this serves to fix the network as a set of connected but ultimately stable territories.

Helen has been active in a campaign to encourage gardeners to maintain fallen and decaying trees and to construct stag beetle nest boxes and log piles. These aesthetic garden features are located on the edges of stag beetle hotspots and are designed to accommodate beetle larvae. She tours South East England seeding deadwood with larvae bred in her own garage. Accessing and connecting the fragmented, private geography of suburban gardens takes time, tact, and the frequent use of freely available satellite imagery. She hopes to connect the metapopulation in a geography of stag beetle flight lines. Her enthusiasms for a more connected landscape resonate with ongoing work to enhance the permeability of domestic gardens—persuading territorial gardeners to tolerate the passage of desired species (like hedgehogs) by cutting holes in fences and creating habitat.[42]

We can find ample evidence of enthusiasms for connectivity in the other examples featured in this book. For example, the RSPB has successfully reintroduced the corncrake to their reserve at the Nene Washes in Cambridgeshire. This involved translocating corncrakes from abundant populations in Poland and rerouting (or perhaps rerooting) their navigation to ensure their annual return. The reserve is subjected to a recalibrated version of the low-intensity grazing and cutting regime pioneered in the Hebrides. In a more radical plan, Vera and his colleagues at SSB envision OVP as a node within a wider national and international ecological network for rewilding.[43] They plan to connect the reserve to a neighboring site and then onto the Veluwe forest—securing an "Oostvaardarswold" that would release the pressure on the large herbivores and extend their grazing dynamics to a wider landscape. This would involve land acquisition and the construction of "ecoducts"—bridges for wildlife over roads and canals. This new geography seeks to reconcile large animals amid densely populated, largely private, and highly productive landscapes. After some significant progress with land acquisition, the project has stalled after falling out of favor with the current agricultural minister.

Although the network topology that informs these interventions might seem intuitive in an age of the Internet and globalization, connectivity has proved difficult to measure. In spite of great efforts by landscape ecologists, there is currently no single universal metric and a growing sense that the search for such an abstraction is futile. When applied to the full diversity of the nonhuman world, the degree of connectivity depends on the scale of analysis, the character of what is being connected, and the methodological difficulties associated with measuring nonhuman mobilities, to name but a few of the challenges encountered. Furthermore, as I explain in more detail, there are frequent conflicts and incommensurabilities between the connectivity requirements of different forms of wildlife and generative ecological, hydrological, and climatological processes.

Efforts to conceive of and enhance connectivity have been influenced by longer-standing anxieties about the undesired mobilities of invasive "alien," "exotic," or "nonnative" species. In the prevalent, but increasingly contested, understanding, these are organisms that have evolved elsewhere, arriving as a result of human action (deliberate or inadvertent), and are able to invade new territory—generally, at the expense of longer-established wildlife. The inconsistent geographies of what constitutes nativity and invasion in this field often tell us more about cultural than ecological problems with nonnative difference.[44] But these anxieties temper conservationists' enthusiasms for unfettered connection and translocation. As a range of analyses have shown, biosecurity policy in the face of invasive species (and other undesired nonhuman mobilities) often requires securing borders—

strengthening a regional topology (the nation, the farm, the nature reserve) or sometimes a fixed network topology—by cutting off connections and their vectors of difference.[45]

This frequently proves practically difficult and can have pathological consequences when conditions of isolation foreclose on resilience or the fluidity necessary for adaptation. For example, some degree of spatial invasiveness is necessary for basic ecological function, not to mention adaptation to a changing climate. Securing UK biodiversity around the territory of the Nation and a present national flora and fauna would be disastrous for future ecologies, shutting off vectors for change, deliberate introduction, or wilding. In thinking through the spatial biopolitics of landscape fluidity, we must attend to the rate, scale, and synchrony of networked movements. Ensuring the persistence of a desired landscape fluidity involves choreographing connectivity, attuning to the diverse topologies of its target forms and processes.

We can find compelling discussion of the challenges associated with this choreography in a recent study of the conservation of the "recombinant ecologies" of a canal in the United Kingdom. Here, Vicky Mason explores tensions in the management of connectivity on these linear landscape features.[46] Canals have emerged as archetypal ecological corridors, neglected linear spaces whose novel ecosystems accommodate diverse valued wildlife and facilitate aquatic, terrestrial, and aerial nonhuman mobilities. Fish swim, insects hop, plants propagate, and bats swarm. Diverse linear ecologies are created, and landscape connections are performed. In cities and in the countryside, people moving on or alongside the canal connect with the nonhuman world in valued ways. But the hydrology of canals makes them risky vectors, for canals overlay and reconnect the discrete boundaries of river catchments. In Mason's account the canal enables aquatic species invasion—easing the mobilities of mammals like mink and a number of fast-growing plants—with actual and anticipated consequences for existing wildlife.

On this canal and elsewhere, current practice is pragmatic and acknowledges the impossibility of eradicating all nonnatives. Securing the vitality of canal ecologies involves working within financial constraints to learn to live with the dynamics of novel ecosystems to avert future homogenization or the eradication of difference. In her critical analysis of this choreography, Mason differentiates between approaches to managing connectivity/invasion premised on a general map of "anticipated" species movement and those that take time to "apprehend" the specific ecological dynamics of this local ecology. In her discussions of mink governance, she argues that preemptive efforts to suppress invasion can jeopardize the fluidity and resilience of a novel ecosystem. A transcendent figure of the biogeography of mink renders them killable, ignoring the specificities of their local patterns of inhabitation. In topological terms, there is a risk of fixing the network in this model of biosecurity. As with the greenwashing of living roofs as green infrastructure, there is risk that the canal becomes a reticular territory, foreclosing on desired fluid dynamics. Her work and those of others in this field suggest that in some instances mink and other global swarmers can accommodate and help nurture biological difference.[47]

Mason's field science of apprehension resonates with the experimental practices of learning to be affected discussed in chapter 2 and the forms of "knowing around wildlife" that I document at OVP in chapter 5. This approach is very much in keeping with the topological approach to conservation and biosecurity advocated by Steve Hinchliffe and his coauthors.[48] Thinking about nonhuman biopolitics topologically requires attuning to the multiplicity of human and nonhuman geographies that animate any given place. Previous chapters show how elephants, corncrakes, and cows all have geographies. These adhere to diverse space–time rhythms and perform diverse topologies. They are caught within and live around prevalent spatial formations of wildlife exchange. Likewise, cycles of carbon, water, hydrogen, and other minerals have inhuman topologies vital for the future of life. The ubiquity and promiscuity of these materials radically undermine a regional topology of discrete territorial units or a hierarchy of nested spatial scales.

CONCLUSIONS

An awareness of this multiplicity of topologies, including those performed by nonhumans, helps inform conservation after the Anthropocene. It does so in a number of ways. First, it helps identify and challenge the territorial trap into which conservation fell in its early efforts to delineate and defend a static nature confined within "island" nature reserves. Thinking about networks then helps map the emerging enthusiasms for and anxieties about connectivity in conservation. Tracing networks maps the geographies of intersecting lines through which landscapes are to be reanimated and by which their difference is threatened. Such lines link between and serve to render porous island territories; they link reserves with the matrix, cities with the countryside, and nations to other nations. Here, we see wildlife on the move across familiar human geographies. But thinking about lines and networks as fixed has risks. A solidified network of prescribed movement can drain ecologies of their vitality—fixing identities to permit their safe passage or rendering life eternal in the greenwashed green infrastructure of prevalent approaches to green urbanism.

Instead, I advocate on behalf of a fluid topology for wildlife. A fluid topology creates space for change; it multiplies and pluralizes what counts as a valuable ecology. A fluid topology of wildlife in the city permits the unstable and unaesthetic ecologies of ruderal brownfield land or saproxylic deadwood. A fluid topology of wildlife enables change within networked mobilities, allowing novel ecologies to become different as they adapt and move. But in advocating fluidity, I am playing with fire. Fluidity is about continuity and gradual change that enables the persistence of valued characteristics in an adapting, evolving assemblage. In permitting change, it replaces a qualitative politics of stability against degradation with a quantitative politics of rates, magnitudes, and rhythms. Fire is the extreme, a topological form defined by persistent creative destruction.

To choreograph landscape dynamics and to seek to modulate such rates of change, conservationists must learn to apprehend and work with diverse nonhuman topologies. An attention to animals' geographies—thinking like an elephant, an insect, or even a molecule—can help attune to the diverse ways in which nonhuman life inhabits the novel ecosystems of an Anthropocene planet. In Ingold's terms, it presents ecology "as the study of the life of lines"[49]—the diverse becomings through which life evolves. This work has the potential to help design better conservation territories—tracking, facilitating, and even disciplining species movements to assist with cohabitation. Thinking like nonhumans helps identify interspecies dependencies and conflicts. It makes manifest the consequences of differential mobilities and degrees of landscape permeability in the face of a changing climate, enthusiasms for trans- and relocation, and the ongoing biotic space–time compression associated with globalization. Finally and perhaps most significant, mapping nonhuman topologies opens experience to the rich diversity of more-than-human ways of being on this planet—or what might more aptly be termed *nonhuman mobilities.*

CONCLUSION

Cosmopolitics for Wildlife

This book is engaged in "anticipatory semantics" with the nascent concept of the Anthropocene.[1] It seeks to shape its emergence and leverage its conceptual and political potential to summon new modes of environmentalism. It wagers that the diagnosis of the Anthropocene offers a shock to thought, a catalyst for new modes of conservation. This is a bold, perhaps forlorn, hope. In the preceding chapters I seek to evaluate, critique, and affirm contemporary forms of conservation to flesh out an alternative mode of thought and practice. I term this a *cosmopolitics for wildlife.* It will be clear by now that this differs markedly from the popular approaches to environmentalism currently energized by the diagnosis of this epoch.

As I explain in the introduction and illustrate through various examples in the preceding chapters, Anthropocene environmentalism has tended toward either greater mastery or forms of naturalism: a final control of or a return to Nature. In chapter 6 and in other parts of my argument, I explore how this cosmopolitics diverges from the popular approach in animal advocacy that elevates a privileged subset of animals to the status of humans by virtue of their approximation to human anatomical, physiological, and cognitive norms.

A cosmopolitics for wildlife departs from the trinity of popular approaches to human–environment relations sketched in Figure 15. Life in the Anthropocene is too strange to be human and afforded rights. It is too social and multiple to be objectified and given a price. And it is too feral to be pure or risky to be liberated in the wilderness. Wildlife conservation after the Anthropocene demands new forms of interspecies responsibility. It needs new types of science and relations between science and politics. It must be founded on different forms of

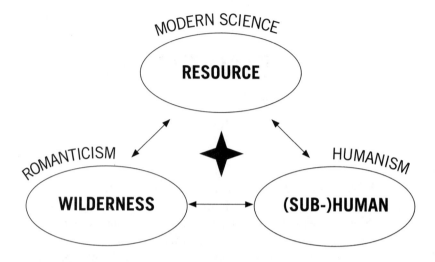

FIGURE 15. A trinity of popular approaches to human–environment relations.

value. In this concluding chapter, I summarize what this book offers to conservation and environmentalism more generally before reflecting on some key points of tension and potential challenges to the approach I offer.

COSMOPOLITICS FOR WILDLIFE

This book seeks to nurture conceptual and practical spaces for wildlife. It finds in wildlife a new ontology for conservation. As I detail in chapter 1, wildlife is not Nature. Wildlife is everywhere. It is among us—in our bodies, our homes, and our cities, as well as in the familiar territories that concern conservationists. Wildlife is shaped by and vital to our actions. Like the elephants in Sri Lanka or the corncrakes in the Hebrides, it is hybrid. It is no good thinking of wildlife out there. We must start living well with wildlife in here. Wildlife has agency—or biopower—shaping human cultural, economic, and political practices.

Wildlife has pasts, presents, and futures. These are not neatly aligned, linear, or concordant. There is no single, universal baseline to which conservation can appeal or toward which restoration can aim. As seen between the Hebrides and the Oostvaardersplassen, divergent future ecologies emerge when conservationists shift from composition to rewilding. Wildlife is thus multinatural. It is immanent. It is difference—where difference is intensive, concerned less with the diversity of current forms and more with the unruly potential to become otherwise. This difference matters. Without it ecologies would cease to function.

The generative potential of nonhuman difference is a vital source of value—driving processes of production, decomposition, sequestration, and aesthetics. As seen in chapter 7, these are increasingly (though awkwardly) commodified—for example, as ecosystem services. Unfettered difference poses significant risks, though, to human life and property. This is the nature dyed red in tooth, claw, and proboscis, our modern predecessors so successfully tamed.

The difference of wildlife inflects the knowledge practices of conservation. Wildlife can be known in many different ways. It is multinatural in a second sense. Different people (and other organisms) learn to be affected by wildlife through a multitude of multispecies entanglements. These have distinct geographies and diverse temporalities. In chapter 2, I detail how the architecture of the human body, enabled by sensory technologies, comes to shape the scope of ecological knowledge, endowing certain species with charisma. In this chapter and in chapter 6, I explore how public sensibilities toward wildlife can be shaped by affective logics of interspecies encounters, including those performed by moving imagery. In chapter 7, I show how such knowledges and sensibilities can conflict in the context of unequal and divergent relations between different publics and charismatic elephants.

The knowledge practices of wildlife conservation are not rational—in the modern sense that they stem from a disembodied calculation of instrumental value. Reason is a rare achievement, uncommon even among conservation biologists. Laid bare, this is not a Science of facts and objectivity. The conservationists we encounter in this book are emotional beings. They care, they love, they hope, they wonder, and they fear. They are not at heart impelled by the anthropocentric logics of ecosystem services. Knowing wildlife is a passionate and embodied practice, and these passions for wildlife generate value in conservation.

I explore different social and economic contexts in which such encounters get valued. I compare the value associated with the epistemic communities of scientists, other naturalists, and experimental filmmakers with the commodified value associated with fee-paying volunteers, private ecotourists, and popular film. Favoring the former, I suggest that the profit imperative of commodified encounters tends to foreclose on difference and the epistemic and ethical possibilities of curiosity.

I argue that knowing wildlife well requires such curiosity and the open-ended care for difference it engenders. I argue that fieldwork necessitates a sensitivity to difference and an openness to surprises. In chapters 4 and 5, we see how field ecologists learn to be affected by corncrakes, cows, and their wider ecologies. In the Hebrides curiosity is directed toward the preservation of a species and the machair ecology. It is oriented around an Experiment, testing a hypothesis deduced from an ecological archetype. At OVP and in some of the disconcerting films we encounter in chapter 6, curious scientists and artists are less sure about what they might find, or even what they might be looking for. Here, a focal organism or ecology becomes an epistemic wild thing, a more-than-human assemblage generative of new ways of thinking and relating. This disposition is rare, perhaps even indulgent, in conservation. But as I demonstrate in the Dutch example, it can result in promising new ways of knowing and relating to wildlife.

Knowledges of wildlife are not all equal, sufficient, or definitive. Multinaturalism is not relativism. There are multiple truths in the accounts that color this book, emergent from the diverse epistemic communities that characterize conservation. Some knowledges are more robust than others in terms of their ability to anticipate ecological change. As I seek to demonstrate in chapters 4 and 5, it is important that we can interrogate their construction, tracing the processes through which key actors come to speak for the environment to critically evaluate the processes of representation and attune to the possibilities of betrayal, where life does not conform. It is also vital that we attend to the politics through which certain knowledges become established, sedimented as orthodox forms of governance.

The difference of wildlife generates such a politics. When multiple futures are possible, no emergent ecology is Natural. Any ecology will favor the livelihoods of some over others, and the identities of these actors are not exclusively human or even organisms. As seen in chapter 5, ecological processes also enact political agency. Furthermore, the shaping power of humans—be they conservationists or other aspirational environmental managers—is constrained. Wildlife has biopower.

This is cosmopolitics—a politics of dynamic processes, diverse agencies, and conditional, contingent, and unstable outcomes. In this book I figure this cosmopolitics as biopolitics—reworking popular theories of biopolitics to acknowledge the biopower of wildlife. Biopolitics as cosmopolitics involves living with nonhumans—conducting, ignoring, subsuming, and being overwhelmed by vital tendencies not subject to human reason.

I demonstrate how conservation is performative, enacting and shaping landscapes and species in relation to powerful representations. These are oligoptical, despite the panoptic aspirations of biodiversity. Conservation emanates from a historic assemblage of standards, technologies, territories, practices, and affective logics that comes to groove present conduct. There is inertia here. Conservation involves ontological choreography—a dance of relations conducted, but not composed, by human actors.

In interrogating the biopolitics of conservation, I figure it as securitization: securing the future of valued forms of life and its diversity through strategic interventions in order to discipline individual subjects and wider populations. Variegated forms of conservation and other modes of nonhuman biopolitics conflict. There is an ontological politics at play that I draw out in more detail in this conclusion. Conceiving of conservation as indeterminate, performative simulation dispels appeals to an authentic Nature and foregrounds the politics of multinaturalism.

The processes associated with this more-than-human politics can be fraught and are frequently unequal. Even among human actors, the possibility of participation and the distribution of potential benefits are skewed. Although this book does not offer a systematic framework for engaging with the politics of conservation, the case studies illustrate diverse means through which conservation publics have come into being and have or have not been included in decision-making processes. At extremes we could contrast the largely consensual and egalitarian collaboration between Hebridean crofters and corncrake conservationists with the asymmetries of Asian elephant conservation in Assam. There are important political ecologies at play here.

Outside of these human actors, the emergent ecologies of the Anthropocene either conducted by conservationists or decimated by other environmental changes would seem to favor certain nonhumans. This is political. We (and it is a narrow we) bear some responsibility for the violence of these extinctions, the losses of forms and possibilities for difference. We bear responsibility for those we save and the conditions under which their salvation is achieved and perpetuated—especially under conditions of captivity.

The ethos of responsibility that finds expression in a cosmopolitics for wildlife is founded on a commitment to the flourishing of significant forms of difference.[2] The fundamental normative premise of this approach is that prevalent modes of environmental and animal management curtail such flourishing—mistreating animals, extinguishing life-forms, and suppressing ecological processes—with severe consequences for animal, environmental, and sometimes human health.

Some modes of conservation are also complicit with this endeavor, managing ecologies toward identities and archetypes that are anachronistic and detrimental to the future health of a warming, unstable, and globalized planet. Flourishing provides an affirmative touchstone. It sounds good, but it is not a simple panacea. It must be worked through in the specifics of the focal political ecology, "staying with the trouble," in Haraway's terms,[3] without making recourse to the generalities of Nature. As such, a more-than-human commitment to flourishing after the Anthropocene raises a series of ontological, biopolitical, and epistemological tensions. Their character helps illustrate both the potential offered by and the challenges facing the cosmopolitics for wildlife I am developing.

TROUBLE IN THE *RAMBUNCTIOUS GARDEN*

In her influential account of new forms of conservation science and management emerging in the Anthropocene, the science writer Emma Marris develops the metaphor of the garden.[4] She suggests that the novel ecosystems of the Anthropocene might be usefully conceived as a "rambunctious" garden: an exuberant and unruly set of ecologies, tended and neglected by people but operating beyond their control. This is a garden in which we live and for which we have responsibility. It is not Eden, nor is it especially pastoral. It is urbanized, industrialized, and technological.

Cast in this fashion, I think there is some potential to this metaphor, which has a long genealogy in environmental thought.[5] The adjective *rambunctious* also animates Haraway's thinking[6] and resonates with the ontology of wildlife I develop in this book. We can develop this metaphor by attending to the cosmopolitics of rambunctious gardening—a theme that is rather absent from Marris's account. Here, I reflect on three broad points of tension. These relate to the forms of difference conservation might let flourish, the security of the human amid the immanence of wildlife, and the relationships between wildlife conservation and capitalism.

Forms of Difference

In chapter 1, I propose wildlife as an alternative ontology for conservation after Nature. I discuss wildlife in the singular, but as the arguments in this book show, it is clear that wildlife is a multiplicity, in Anne-Marie Mol's terms. It subsumes more than one but less than many ontologies. In the preceding chapters, we encounter at least five ways of cutting up wildlife for conservation, not all of which are concordant.

In chapter 3, I explain that orthodox conservation has tended to focus on *species*—aggregations of organisms with morphological similarities. Species provide the prevalent, even intuitive, ontology for much contemporary conservation. A taxonomy of species guides the panoptic aspirations of biodiversity and the CBD. It informs the IUCN's calculations of rarity and threat. It shapes the scientific and legal infrastructure and modes of practical management encountered in chapter 4. A subset of charismatic species is the focus of the spectacular encounters associated with ecotourism and natural history film.

In contrast, conservationists at OVP understand species (in this case, cows and horses) as expressions of ecological *processes* and tools for the delivery of ecological functions—grazing, predation, and decomposition, for example. Here, the integrity and even persistence of a species is secondary to the flourishing of an ecology. While the machair landscapes of the Hebrides are managed for corncrakes, the wild cows of OVP are introduced to manage a dynamic landscape. Animal advocates concerned with the welfare of the large herbivores at OVP and ecotourists and filmmakers after touching encounters focus on the *individual* animals that comprise a species or herd. Here, the suffering of an individual is primary to the flourishing of a species or its ecology.

Furthermore, ethologists argue that these dedomesticated animals have learned to be different. Individually and as a group, they have developed new bovine and equine *cultures*: behaviors, sociabilities, physiologies, and immunities worthy of conservation value. Finally, the difference of wildlife is reconfigured when we attend to *genetics,* which in its most radical formulation in the context of microbes fundamentally undermines the familiar logics of organisms and the aggregate and individual welfare.

These five ways of cutting up wildlife result in different modes of conservation. We could generate many more if we attended to the emerging abstractions of carbon or the metrics for calculating ecosystem services. While the interests of different targets can often be aligned, there are frequent, important, and sometimes discordant ontological politics at their interfaces.

These tensions are perhaps most clearly expressed in the differentiation and conservation of Asian elephants, which I discuss in chapter 1. For example, much global Asian elephant conservation is geared toward securing *Elephas maximus.* This cut and its performative assemblage prioritizes the species—a group of organisms that could theoretically breed but are Globally Threatened. But *Elephas maximus* subsumes an unstable number of subspecies. "Splitters" and "lumpers" among elephant taxonomists currently claim between two and five varieties, differentiated corporeally by their genes and (to a lesser extent) morphology.[7] Their contested taxonomies disaggregate *Elephas maximus* into "evolutionary significant units"[8] of intraspecies difference.

These divisions territorialize the ageographic ontology of the species to differentiate by biogeography and thus phylogeny—or evolutionary history. Here, groups of elephants are valued for the distinctive pasts they embody and their potential for future adaptive differentiation. As a consequence, evolved intraspecific genetic variation like that embodied within the Asian elephants on Borneo is afforded much higher priority than the recent geographic isolation of feral populations of former logging elephants abandoned on the Andaman Islands.[9] Conservationists concerned with intraspecific variation resist proposals to consolidate the metapopulation of *Elephas maximus* in a smaller number of reserves to secure the species' survival.

These ranked classifications might work for Natural animals existing before the Anthropocene. But they are complicated by the forms of pachyderm difference that emerge from the entanglement of Asian elephants as companion species in human histories, ecologies, and affections. For example, the phenotypic plasticity, corporeal generosity, and hybrid ecologies of Sri Lanka's feral elephant population pose far-reaching challenges to species- and genetics-based modes of classifying elephant difference. If we accept that the immigrant escapees, abused rogues, and free-ranging matriarchs encountered in chapter 1 express multiple elephant cultures, then other forms of difference (and associated modes of companionship) become significant. This is even more the case when the focus turns to captive elephants, like those encountered in chapter 6. These animals sustain the lucrative economies of ecotourism as well as traditional cultural practices. Their conservation requires continued captivity maintaining animals isolated from the breeding metapopulation, habituated to solitary confinement and poor welfare. These animals suffer for their captivation.

Asian elephants are rarely discussed as ecological agents, although they play a vital role as instigators of disturbance as grazers and seed dispersers. An ontology of ecological processes is more commonly used to value the difference of African elephants. Excess animals are culled in parts of southern Africa when their populations exceed a calculated carrying capacity and threaten the integrity of a desired ecology. This creates tensions with those concerned with species and individual welfare.[10]

African elephants have emerged as focal species for rewilding. Like the cows and horses at OVP, they have been proposed as surrogates for extinct antecedents and their ecological functions. Reintroductions have been proposed in North America and the United Kingdom.[11]

African elephants figure as surrogate mothers in plans to re-create the mammoth for reintroduction in Siberia.[12] Further schemes imagine African elephants instigating trophic order in the novel ecosystems of parts of rural Australia, grazing invasive grasses familiar to their African ecologies.[13] These ideas currently seem farfetched and un-likely, but they have prompted debate and opposition among those concerned with the health and welfare of individuals, species, and wider ecologies.[14]

This vignette illustrates five significant forms of nonhuman differ-ence with strong arguments and active and sometimes conflicting programs in support of their future flourishing. In the critical and affirmative analysis that cuts across the variegated examples that make up this book, I err toward a processural ontology of wildlife and an associated focus on flourishing. Here, the future of a species and the welfare of its constitutive individuals matter, especially when they align with the conservation of the heterogeneity of intensive forms of difference necessary for securing the resilience or transformative capacity of a focal ecology. The composition of organisms, species, genes, and habitats offers a redoubt—strategic essentialisms in the face of destructive fluidity—but we should not let them fix the generative processes that give inhabited ecologies their vitality and health. I thus acknowledge that there are legitimate reasons for conservationists to subject animals to pain and death and even to let animal cultures and species go extinct. I am sympathetic to the arguments for prioritizing individual animal suffering and contesting the abject biopolitics of much modern agriculture. I find the commodified degenerative re-lations of captivity extremely problematic, whatever their purported conservation benefits.

This is complicated terrain requiring a fuller analysis than this space allows. As Donna Haraway and Kathryn Yusoff remind us, making something killable and witnessing extinction are processes demanding ethical responsibility.[15] Conservationists should and, in my opinion, generally do take these matters seriously. My point here is that for conservation after the Anthropocene the choice of units for measuring difference matters. Conservation is engaged in ontological choreography in concert with nonhumans. There are multiple ways of cutting up that difference, none of which are Natural, all of which will have political consequences.

Wildlife Conservation as Biosecurity

Theory, fiction, and futurology adopting an extreme version of the fluid, processural ontology of wildlife I favor in this book frequently move from a now commonplace decentering of the human toward its dissolution or disavowal. In dystopic imaginations wildlife is left to run amok amid the ruins of civilization, enacting vengeance for human abuse on a diminished, fearful, and subservient population. Contrasting hyperbolic visions see the human becoming lost through the distribution of intelligence and the biotechnological blurring of human–animal and human–machine boundaries.[16]

These twin strands of posthumanist thinking often suggest that there is a tension (or even a paradox) in the conjunction of the words *wildlife* and *conservation*. For example, in an engagement with recent work on vital materialism and environmental thought, the geographer Bruce Braun argues that "eco-politics must be oriented not toward conservation, since the world never holds still, but to the possibilities and consequences of a 'new earth' and a 'new humanity' that is still to come."[17]

Taken this way and excusing the rather binary future offered, wildlife conservation becomes something of a contradiction in terms, the implicit normativity being an admonishment of actions that constrain the lively potential of evolving ecologies. This reading would perhaps be unfair, and elsewhere, Braun cautions against any naïve celebration of vital powers and fluidity,[18] but it resonates with forms of environmental thought informed by Deleuze, Nietzsche, and vitalist philosophers. This book is more cautious, even conservative, in tone. I maintain that wildlife needs conservation. It needs science. It needs technology. It needs administration. It needs politics. The forms of conservation I envisage are biopolitical or, better still, cosmopolitical. Although they are posthumanist, they maintain a latent critical humanist concern with protecting valued forms of life and subjectivity—both human and nonhuman. Anything else would be a derogation of the stewardship responsibilities inherent to the Anthropocene or a perpetuation of the unequal historical distribution of the costs and benefits of global biodiversity loss. Marginal social groups consistently get dispossessed, diseased, and eaten when wildlife is let loose.

In this reading, wildlife conservation is akin to a mode of biosecurity. It seeks to secure a desired future bio through nurturing relations and cultivating abnegations in order to enable companionship between humans and other species. In so doing, it secures a figure of the human—enacting modes of environmentality in the guise of good environmental citizenship. As Steve Hinchliffe and his coauthors argue, a cosmopolitical approach to wildlife requires a model of biosecurity different from that which characterizes much contemporary conservation (and other forms of environmental governance).[19] It demands a figure of the human more nuanced and differentiated than the rational economic actor favored by the dominant modes of social science that inform conservation policy. It necessitates an experimental mode of field science open to surprises. And it cannot be premised on anticipating and securing fixed identities and territories. As discussed in chapters 5 and 8, such an approach risks rendering the present eternal.

Instead, as Hinchliffe and his coauthors compellingly illustrate in their work on urban conservation and the management of emergent zoonotic disease, biosecurity (of both the human and the nonhuman bio) requires a speculative, topological approach. This amounts to a probiotic mode of biopolitics. It involves working with the biopower of the nonhuman world. Such entanglements must be attuned to the inherent vitality and promiscuity of wildlife. They should be "future-invocative," sensing, nurturing, guiding, stalling, and foreclosing on emergent processes.[20] They must channel the trajectories of fluid ecologies while playing with fire and the generative potentials of disturbance. I find the most heartening evidence of this approach in the management of OVP, in elements of urban conservation, and in curious modes of experimental film.

To make this case, I have to read them against the grain, but I think there is great potential in wilding as a more open mode of conservation. Shorn of the prefix re-, it opens up the temporalities of conservation. It is about the future as much as the past. It offers hope.

In short, it offers a form of conservation that is not about retreat. In Bruno Latour's terms, this model turns to face the future, learning to live with and love the "monsters" that characterize the Anthropocene.[21] Conservation remains a biopolitical practice: living well with wildlife requires fences, rifles, and cameras alongside legal, economic, and political technologies for mediating and deliberating companionship. It takes place in the open, without recourse to a lost Nature or a universal Human.

Wildlife and Capitalism

An eternal Nature has been a staple of environmental politics—on both the left and the right. It has been deployed frequently to contest the operations (or at least the excesses) of capitalism, the instrumental use of the nonhuman world as resources, and other undesired modern developments. A pristine or premodern Nature provides a redoubt—a stable source of value and locus for unalienated human–environment relations. I am arguing that this has to stop. But giving up on Nature in place of a hybrid, immanent, and multinatural world is risky. Some would say this is irresponsible, complicit, or defeatist, especially given the global ascendency of neoliberal capitalism and the rise of free-market environmentalisms.[22]

Indeed, the growing encroachment of neoliberal modes of environmental governance upon policy domains formerly characterized by their opposition to capitalism has been enabled by the recasting of ecologies as fluid, fungible, and generative assemblages that can be subsumed to the logics of "ecosystem service" provision. Here, natures can be abstracted, commodified, traded, and enhanced through financialization—secured by making them subject to the governmental and biopolitical technologies previously associated with financial markets dedicated to their exploitation.[23] Discursively, this process has involved normalizing markets and naturalizing the innovative character of capitalism through the metaphorical alignment of biological and economic processes and practices.[24] If a fixed Nature is required for authoritarian modes of conservation premised on centralized state power, so a fluid, individualistic, and fungible nature is necessary for neoliberalism.

The rise of neoliberal models and associated natures is increasingly true of conservation. The commodification of encounter value in ecotourism analyzed in chapter 7 tracks parallel innovations in the financialization of the reproductive value of ecologies. These include the invention and introduction of markets for biodiversity offsets, species banking, bioprospecting, and myriad forms of carbon sequestration, such as Reducing Emissions from Deforestation and Forest Degradation (REDD). These developments are alarming and have been subject to a growing body of critical work that lays bare their social and ecological inadequacies as modes of environment governance.[25] As with the scope and conduct of ecotourism premised on the commodification of encounter value, there are specific types of natures (both goods and services) that capital can see (to use Morgan Robertson's phrase[26]) and forms of environmental citizenship that are valued. As the global conservation industry powers up such financialized models of environmental governance, so they will growing performative effects on life after the Anthropocene.

I should be clear that although the hybrid, immanent, and experimental modes of conservation I advocate sacrifice the oppositional power of Nature, they do not equate with the future natures that advocates imagine will be (geo- and bio-) engineered and secured through neoliberal environmentalism—first and perhaps most important, because wildlife is public, the property of a more-than-human citizenry unable or undisposed to participate in relations of commodified consumption. The unnatural wilds of the Anthropocene should be understood as multispecies commons, claimed and contested by diverse economic interests.

Second, wildlife is unruly. Its rambunctious potential unsettles the abstracted and fixed categories associated with emergent forms of species banking and the curtailed and anthropocentric immanence of ecosystem services.[27] Finally, wildlife is multinatural in the forms of value it generates. Charismatic wildlife in all its rich variety exceeds the narrow logics of exchange value. Yes, it can be commodified. Yes, it offers nonconsumptive resources for economic growth, but these are subtended by a rich and multicultural diversity of curiosities. Multinatural wildlife necessitates a knowledge politics open to epistemological discord, not a politics driven by market power.

A cosmopolitics of wildlife therefore necessitates a politics and a form of political economy in opposition to prevalent forms of neoliberal environmentalism, in which future natures are secured for the interests of a limited and contemporary human few. We encounter fragments of such a configuration in the various examples and relations that illustrate this book. A cosmopolitical ecology for wildlife requires a multispecies model of justice that eludes the trinity of popular approaches diagrammed in the opening of this chapter. This will be unfamiliar and perhaps uncomfortable to existing approaches to the political ecology of conservation. Distributive justice for wildlife conservation involves a more-than-human citizenry linked by affective relations that exceed those of production and exchange. It features subjects, forms of value, and affective logics beyond the narrow confines of human reason or interpersonal sympathies. It requires differentiated senses of responsibility to distant human and nonhuman others.

In many cases the Anthropocene is a disaster, but it also presents opportunities. Cast off from Nature, conservation can begin to think differently. In wildlife I wager that a world of multiple natures offers more interesting, hopeful, and democratic futures for environmentalism.

ACKNOWLEDGMENTS

This book first took shape in the School of Geographical Sciences at the University of Bristol, where I did my undergraduate and graduate studies. It has followed me to other institutions, including Kings College London, the University of Oxford, and (more briefly) the University of British Columbia and the University of Peradeniya. Along the way it has benefited from conversations with many intelligent, generous, and original thinkers.

Sarah Whatmore first got me thinking about Nature when I was a teenager. She helped me spend a summer following corncrakes in the Hebrides and has since been a wonderful mentor, collaborator, and boss. I am grateful for her advice and support over the past fifteen years. Sarah's inspiration has been nurtured through my engagements with a growing network of more-than-human geographers (not all of whom would accept this title). I thank Simon Naylor, Paul Cloke, and J. D. Dewsbury for helping shape my PhD. My thinking has benefitted greatly from conversations and textual encounters with Bill Adams, Ben Anderson, Trevor Barnes, Andrew Barry, Maan Barua, Jane Bennett, Bruce Braun, Irus Braverman, Dan Brockington, Raymond Bryant, Henry Buller, Noel Castree, Chloe Choi, Gail Davies, David Demeritt, Rebecca Ellis, Rob Francis, Mike Goodman, Beth Greenhough, Lesley Head, Clare Herrick, Steve Hinchliffe, Tim Hodgetts, Jim Igoe, Paul Jepson, Alex Loftus, Emma Marris, Derek McCormack, Mara Miele, Kelsi Nagy, Mark Pelling, Mike Raco, Emma Roe, Krithika Srinivasan, Sian Sullivan, Clare Waterton, and Kathryn Yusoff.

Since 2010 I have enjoyed working with Clemens Driessen. Clemens is partly responsible for the thinking on rewilding in chapter 5. Thanks go to him for his help and for opening my eyes to what is different about continental Europe. The Economic and Social Research Council funded the majority of this book's research, and I am grateful for their doctoral and postdoctoral support. Thanks go to Jason Weidemann—my editor—and the team at the University of Minnesota Press.

None of this research would have been possible without the generosity of the many conservationists who gave me their time and access to their projects, patiently explained their thinking, and tolerated my social science. Their passion, curiosity, and dedication are the inspiration for this book. I hope it gives something in return.

NOTES

INTRODUCTION

1. The term is generally associated with the Nobel Prize–winning geologist Paul Crutzen (2002). It has still to be formally accepted by the International Commission on Stratigraphy but has been successfully promoted by a small cadre of geologists and "earth system scientists."

2. For a review, see Zalasiewicz et al., "Stratigraphy of the Anthropocene."

3. Take for example a special edition of *The Economist* on May 26, 2011, entitled "Welcome to the Anthropocene."

4. See Steffen et al., "The Anthropocene."

5. This is especially the case in relation to climate change. See Hulme, *Why We Disagree about Climate Change.*

6. Latour, *Politics of Nature*; Latour, *We Have Never Been Modern.*

7. I take the word *multinatural* from Latour, who borrows it from the anthropologist Eduardo Viveiros de Castro.

8. Wapner, *Living through the End of Nature.*

9. The controversial potential of geoengineering is discussed in Hamilton, *Earthmasters.*

10. See for example Lynas, *The God Species*; Brand, *Whole Earth Discipline*; Nordhaus and Shellenberger, *Break Through.*

11. See for example the writings of the Dark Mountain Project on "uncivilisation" at http://dark-mountain.net. This narrative is implicit also within Alan Weisman's best seller *The World without Us.*

12. Clark, *Inhuman Nature.*

13. Ogden et al. *Global Assemblages, Resilience, and Earth Stewardship in the Anthropocene.*

14. For an example of this discourse, see Sachs, *Common Wealth.* For nuanced and differing critiques of this universalizing tendency, see Chakrabarty, *The Climate of History*; Žižek, *Living in the End Times.*

15. Plumwood, *Environmental Culture.*

16. This point is made in passing in Haraway, *When Species Meet*. See also Nagy, *Trash Animals*.

17. Žižek, *In Defense of Lost Causes*; Swyngedouw, "Apocalypse Forever?"

18. Steffen, "The Anthropocene."

19. Ibid., 619.

20. These forms of stewardship are somewhat out of alignment with the provenance of the figure of the environmental steward. This is traced generally to the writings and land ethic of Aldo Leopold—a key advocate for untrammeled wilderness.

21. I take the label *cosmos* from Haraway and Latour, who in different ways are indebted to the philosophy of Isabelle Stengers and her "cosmopolitics." See Latour, *An Attempt at a "Compositionist Manifesto"*; Haraway, *When Species Meet*; Stengers, *Cosmopolitics*.

22. See Rockstrom et al., *A Safe Operating Space for Humanity*. The prospects for a geologically responsible citizenry have recently been the subject of some compelling and important work in geography and geophilosophy. See Clark, *Inhuman Nature*; Woodard, *On an Ungrounded Earth*; Clark et al., *Capitalism and the Earth*; Yusoff, "Geologic Life."

23. For influential accounts of this work, see Thrift, *Non-representational Theory*; Ingold, *The Perception of the Environment*; Stewart, *Ordinary Affects*.

24. I take this term from Eva Hayward and her work on animal film. See Hayward, "Enfolded Vision."

25. Whatmore, *Hybrid Geographies,* esp. chap. 2.

26. This sentiment was most famously articulated in McKibben, *The End of Nature*. It continues to inform writings about the Anthropocene. For example, in a recent paper in a special edition of the *Philosophical Transactions of the Royal Society,* Ellis argues, "The terrestrial biosphere is now predominantly anthropogenic, fundamentally distinct from the wild biosphere of the Holocene and before. From a philosophical point of view, nature is now human nature; there is no more wild nature to be found, just ecosystems in different states of human interaction, differing in wildness and humanness" (1027).

27. This point is famously and forcefully made by Bill Cronon. See Cronon, "The Trouble with Wilderness."

28. See Hinchliffe et al., "Urban Wild Things."

29. See for example Kareiva, "Conservation in the Anthropocene"; Caro et al., "Conservation in the Anthropocene"; Soule, "The New Conservation."

30. Hobbs et al., *Novel Ecosystems*.

31. Marris, *Rambunctious Garden*.

32. Cronon, "Uncommon Ground"; Soule and Lease, *Reinventing Nature?*

33. Sarah Whatmore makes this point in *Hybrid Geographies*.

34. I take this phrase from H. Lorimer, "Cultural Geography."

35. I take this phrase from Latour, *Pandora's Hope*.

36. Ibid.

37. I take this term from Despret, "The Body We Care For."

38. I take this phrase from Carter and McCormack, "Film, Geopolitics, and the Affective Logics of Intervention."

39. The concept was originally outlined in Deleuze and Guattari, *A Thousand Plateaus*. For further elaborations, see Dewsbury, "The Deleuze-Guattarian Assemblage"; Anderson et al., "On Assemblages and Geography"; Bennett, "The Agency of Assemblages."

40. I take this phrase from Whatmore and Braun, *Political Matter*.

41. I take this definition from Hans-Jörg Rheinberger and the development of his work in relation to the field sciences by Matthias Gross. See Rheinberger, *Toward a History of Epistemic Things*; Gross, *Ignorance and Surprise*; Gross, *Inventing Nature*.

42. Callon, Lascoumes and Barthe, *Acting in an Uncertain World*.

43. Bennett, *The Enchantment of Modern Life*.

44. See especially Hinchliffe et al., "Urban Wild Things"; Hinchliffe, "Reconstituting Nature Conservation."

45. Wildlife is the concern of one of the United Kingdom's largest environmental NGOs, the Wildlife Trusts, who have perhaps done the most to champion urban ecologies and urban publics in UK conservation. See the Wildlife Trusts website at www.wildlifetrusts.org.

46. Haraway, *When Species Meet*.

47. For an introduction, see Buscher et al., "Towards a Synthesized Critique of Neoliberal Biodiversity Conservation"; Igoe, Neves, and Brockington, "A Spectacular Eco-tour around the Historic Bloc."

48. Worster, *Nature and the Disorder of History*.

49. See Latour, "Why Has Critique Run out of Steam?"

50. This rather unfortunate label seems to have been coined by Keith Kloor in a 2005 article on *Slate*. Others prefer the label *eco-pragmatist*.

51. At a recent gathering at which she was discussing her book, E. O. Wilson asked Emma Marris where she was planning on "planting the white flag she was carrying"? See the report on Andrew Revkin's *Dot Earth* blog. Andrew Revkin, "Emma Marris: In Defense of Everglades Pythons," *New York Times*, August 17, 2012, http://dotearth.blogs.nytimes.com/2012/08/17/emma-marris-in-defense-of-everglades-pythons/?_php=true&_type=blogs&_r=0.

52. In *When Species Meet*, Haraway explains how she takes this term from the work of Charis Thompson. See Thompson, *Making Parents*. For a comparable approach to the performativity of matter, see Barad, "Posthumanist Performativity."

53. Mol, *Ontological Politics.*

54. Foucault, *Security, Territory, Population,* 1.

55. See especially Foucault, *Society Must Be Defended;* Foucault, *The Birth of Biopolitics*; Foucault, *Security, Territory, Population.* For a more general introduction to biopolitics, see Lemke, *Biopolitics.*

56. Agrawal, *Environmentality.* See also Rutherford, *Governing the Wild;* West, *Conservation Is Our Government Now;* Lowe, *Wild Profusion*; Tsing, *Friction.*

57. For a range of empirical foci for biopolitical analysis, see Rose, *The Politics of Life Itself;* Cooper, *Life as Surplus*; Rajan, *Biocapital*; Dalby, *Biopolitics and Climate Security in the Anthropocene*; Grove, *Insuring "Our Common Future?"*

58. For a discussion, see Darier, *Discourses of the Environment.*

59. For a range of examples, see Shukin, *Animal Capital*; Holloway et al., "Biopower, Genetics, and Livestock Breeding"; Hinchliffe et al., "Biosecurity and the Topologies of Infected Life"; Srinivasan, "The Biopolitics of Animal Being and Welfare"; Franklin, *Dolly Mixtures*; Braverman, *Zooland*; Helmreich, *Alien Ocean.*

60. Agamben, *The Open.*

61. Foucault, *The History of Sexuality.*

62. For example, see Shukin, *Animal Capital.*

63. Wolfe, *Before the Law*; Haraway, *When Species Meet*; Esposito, *Immunitas.*

64. Haraway, *When Species Meet.*

65. Haraway, Hinchliffe, and Whatmore develop their conceptualizations of cosmopolitics from the original work of Isabelle Stengers. See Stengers, *Cosmopolitics*; Stengers, *Cosmopolitics: II.*

66. I draw in particular on the ideas articulated in the following papers: Hinchliffe et al., "Biosecurity and the Topologies of Infected Life"; Hinchliffe, "Reconstituting Nature Conservation"; Hinchliffe et al., "Urban Wild Things"; Hinchliffe and Lavau, "Differentiated Circuits."

67. Hinchliffe, "Reconstituting Nature Conservation."

68. Lane et al., "Doing Flood Risk Science Differently."

69. Bright, *Life Out of Bounds.*

1. WILDLIFE

1. Santiapillai et al., "A Strategy for the Conservation of the Asian Elephant in Sri Lanka."

2. Fernando, "Elephants in Sri Lanka."

3. Fleischer et al., "Phylogeography of the Asian Elephant"; Vidya et

al., "Range-wide Mtdna Phylogeography Yields Insights into the Origins of Asian Elephants."

4. Clutton-Brock, *A Natural History of Domesticated Mammals.*

5. Fernando and Lande, "Molecular Genetic and Behavioral Analysis of Social Organization in the Asian Elephant"; Fernando et al., "Ranging Behavior of the Asian Elephant in Sri Lanka."

6. Fernando et al., "Perceptions and Patterns of Human–Elephant Conflict in Old and New Settlements in Sri Lanka."

7. Sukumar, *The Living Elephants.*

8. Bennett, *Vibrant Matter.*

9. Fernando, "Elephants in Sri Lanka"; Fernando et al., "Perceptions and Patterns of Human–Elephant Conflict in Old and New Settlements in Sri Lanka."

10. These are most clearly expressed in Latour, *We Have Never Been Modern*; Haraway, *Simians, Cyborgs, and Women.*

11. Hybridity talk can even now be heard among the higher echelons of the environmental movement as they start to grapple with the implications of the Anthropocene. See for example a recent essay by Peter Kareiva, the chief scientist for the Nature Conservancy (one of the largest U.S. conservation NGOs), who quips, "One need not be a postmodernist to understand that the concept of Nature, as opposed to the physical and chemical workings of natural systems, has always been a human construction, shaped and designed for human ends." Kareiva, "Conservation in the Anthropocene."

12. Picking up on McKibben's famous, mournful diagnosis of the "end of Nature," Cronon notes how an environmentalism whose valued Nature is premised on human absence "takes to a logical extreme the paradox that was built into wilderness from the beginning: if nature dies because we enter it, then the only way to save nature is to kill ourselves." Cronon, "The Trouble with Wilderness," 83. Cronon's hybrid thinking is not obviously influenced by Haraway or Latour, though Donna Haraway was one of the thinkers gathered by Cronon for the event that led to the famous edited collection *Uncommon Ground,* in which his wilderness essay was published. We can detect Latour's and Haraway's influences on a comparable debunking of wilderness offered by Sarah Whatmore and Lorraine Thorne a few years later. See Whatmore and Thorne, *"Wild(er)ness."*

13. See for example Willis and Birks, *"What Is Natural?"*

14. Hobbs et al., *"Novel Ecosystems."*

15. Ellis and Ramankutty, *Putting People in the Map*; Ellis, *Sustaining Biodiversity and People in the World's Anthropogenic Biomes.*

16. Daily et al., "Countryside Biogeography," 2.

17. Rosenzweig, *Reconciliation Ecology,* 201.

18. Higgs, *Nature by Design*; Keulartz, "The Emergence of Enlightened Anthropocentrism in Ecological Restoration."

19. Marris, *Rambunctious Garden*.

20. See for example the wide range of animals covered in the Reaktion Animal Series as well as the following for a flavor of the diversity of this field: Cassidy and Mullin, *Where the Wild Things Are Now*; Raffles, *Insectopedia*; Helmreich, *Alien Ocean*; Song, *Pigeon Trouble*; Bekoff, *Minding Animals*.

21. Opening up the category *animal* is most commonly associated with writings on animals by Jacques Derrida. See Derrida and Mallet, *The Animal That Therefore I Am*.

22. Haraway, *When Species Meet*.

23. Haraway's intervention has catalyzed and resonated with a range of work tracing modes of companionship across an array of animal and other forms. For a range of examples, see Greenhough, "Where Species Meet and Mingle"; Tsing, "Unruly Edges"; Davies, "Caring for the Multiple and the Multitude"; Bear and Eden, "Thinking Like a Fish."

24. Lorimer, "Elephants as Companion Species."

25. Latour, *We Have Never Been Modern*; Haraway, *When Species Meet*.

26. Little is known about this microbial biodiversity, but there is a growing interest in the postgenomic sciences in mapping its ecology and understanding the consequences of its dynamics. Big science initiatives like the Human Microbiome Project are being driven in large part by the increasing affordability and thus availability of technologies for metagenomic sequencing and the promise of new understandings and products for improving human health. See Turnbaugh et al., "The Human Microbiome Project." To date, there has been limited attention to similarities between the constitution and dynamics of micro- and the macrobiomes and the shifting dynamics of their forms of biodiversity.

27. Early results from the Human Microbiome Project suggest that (1) the total microbial cells found in association with human bodies may exceed the total number of human cells making up our bodies by a factor of ten to one and (2) the total number of genes associated with the human microbiome could exceed the total number of human genes by a factor of one hundred to one. Human Microbiome Project, "Structure, Function and Diversity."

28. Fowler and Mikota, *Biology, Medicine, and Surgery of Elephants*.

29. Rothschild and Laub, *Hyperdisease in the Late Pleistocene*.

30. Sukumar, *The Living Elephants*.

31. Mikota, *Review of Tuberculosis in Captive Elephants and Implications for Wild Populations*.

32. Alexander et al., *Mycobacterium Tuberculosis*.

33. Diprose, *Corporeal Generosity*.

34. Margulis and Sagan, *Acquiring Genomes*, 31.

35. Hird, *The Origins of Sociable Life*.

36. Yusoff, "Geologic Life."

37. This point is elaborated in work by Eric Swyngedouw. See Swyngedouw, "Apocalypse Forever?"

38. For an introduction and short history of the theory, see Latour, *Reassembling the Social*.

39. This term was coined by De Landa in *Intensive Science and Virtual Philosophy*.

40. See for example Latour, "To Modernise or Ecologise?"

41. Whatmore and Braun, *Political Matter*.

42. Kirksey and Helmreich, "The Emergence of Multispecies Ethnography".

43. Haraway, *When Species Meet*.

44. See for example Acampora, *Corporal Compassion*; Greenhough and Roe, "Ethics, Space and Somatic Sensibilities."

45. See for example Lorimer, "Herding Memories of Humans and Animals"; Game, "Riding"; Bear and Eden, "Thinking Like a Fish?"; Lorimer and Whatmore, "After 'The King of Beasts.'"

46. Laland and Bennett, *The Question of Animal Culture*.

47. For an exemplary piece of work, see Fuentes, "Naturalcultural Encounters in Bali."

48. Bradshaw, *Not by Bread Alone,* 145; Bradshaw, *Elephants on the Edge*; Lawrence and Spence, *The Elephant Whisperer*.

49. Bates et al., "Elephants Classify Human Ethnic Groups."

50. I take the term *inhuman* from Nigel Clark. See Clark, *Inhuman Nature*. See also Ansell-Pearson, *Germinal Life*; De Landa, *A Thousand Years of Nonlinear History*; Morton, *The Ecological Thought*.

51. This alignment is best illustrated in Latour's recent collaborations with market environmentalists. See Shellenberger and Nordhaus, *Love Your Monsters*.

52. Clark's concept of geopower resonates with Jane Bennett's popular recent work on "vital matter" in which she flags the agencies (or what she terms the "thing-power") of a host of nonhuman actors, both living and inert, anthropogenic and otherwise. See Bennett, "The Force of Things"; Bennett, *Vibrant Matter*.

53. Deleuze and Guattari, *A Thousand Plateaus*.

54. Massey, *For Space*.

55. See for example De Landa, *Intensive Science and Virtual Philosophy*.

56. See Zimmerer, "The Reworking of Conservation Geographies".

57. Manning et al., "Landscape Fluidity," 193.

58. Ibid., 194.

59. Bengtsson et al., "Reserves, Resilience, and Dynamic Landscapes."

60. Botkin, *Discordant Harmonies*.

61. Sri Lanka Department for Wildlife Conservation, "National Policy for the Conservation and Management of Wild Elephants in Sri Lanka."

62. Fernando et al., "Perceptions and Patterns of Human–Elephant Conflict."

63. Law and Mol, "Situating Technoscience," 614.

64. For a review, see Martin and Secor, "Towards a Post-mathematical Topology."

65. Murdoch, *Post-structuralist Geography*; Hinchliffe et al., "Biosecurity and the Topologies of Infected Life"; Law and Mol, *Situating Technoscience*; Lorimer, "Living Roofs and Brownfield Wildlife."

66. Warren, "Perspectives on the 'Alien' versus 'Native' Species Debate"; Cassidy and Mullin, *Where the Wild Things Are Now*.

67. Hinchliffe and Whatmore, "Living Cities."

68. Clark, "The Demon-Seed."

69. For a general introduction to phylogeography, see Avise, *Phylogeography*. Recent work on elephants can be found in Fleischer et al., "Phylogeography of the Asian Elephant"; Vidya et al., "Range-wide Mtdna Phylogeography Yields Insights into the Origins of Asian Elephants."

70. Jazeel, "'Nature,' Nationhood and the Poetics of Meaning."

71. Fernando et al., "Mitochondrial DNA Variation."

72. See Crooks and Sanjayan, *Connectivity Conservation*. There has been a lot of interest in connectivity in England as a result of the 2010 Lawton Review into Nature Conservation in the country. This was entitled Making Space for Nature.

73. Jongman et al., "European Ecological Networks and Greenways."

74. This is a contested development. See for example Hewitt et al., "Taking Stock of the Assisted Migration Debate"; Ricciardi and Simberloff, "Assisted Colonization Is Not a Viable Conservation Strategy."

75. Rosenzweig, "The Four Questions."

76. I take this phrase from Bright, *Life Out of Bounds*.

77. Olden, "Biotic Homogenization."

78. See for example Davis et al., "Don't Judge Species on their Origins"; Schlaepfer et al., "The Potential Conservation Value of Non-native Species."

79. Sax and Gaines, "Species Diversity."

80. Sukumar, *The Living Elephants*.

81. Menon et al., *Rights of Passage*.

82. I take the term *beastly places* from Philo and Wilbert, *Animal Spaces, Beastly Places*.

83. This approach to place is developed in Massey, *For Space*.

84. I take the term *multiplicity* from Anne-Marie Mol, for whom it describes the possibility of more than one but less than many ways of associating with and enacting material realities. Multiplicity leads to ontological politics. See Mol, *The Body Multiple*. For a useful discussion and illustration of the significance of this approach to environmental politics, see Hinchliffe, *Geographies of Nature*.

85. For example, see Gaston, *Biodiversity*.

86. Deleuze, *Difference and Repetition*, 222.

87. Wilson, *The Diversity of Life*.

88. The concepts of ontological politics and multiplicity are developed in the context of environmental management in Lien and Law, "Emergent Aliens."

2. NONHUMAN CHARISMA

1. Craig is an avatar for an amalgam of corncrake surveyors and scientists I worked with during this research.

2. I take this phrase from Despret, "The Body We Care For: Figures of Anthropo-Zoo-Genesis."

3. For a compelling account of the challenges of surveying birds, this time in an urban environment, see Hinchliffe, "Reconstituting Nature Conservation."

4. This point is elaborated in more detail in Kay Milton, *Loving Nature*.

5. Latour, *Pandora's Hope*.

6. Carter and McCormack, "Film, Geopolitics, and the Affective Logics of Intervention."

7. Lorimer, "Cultural Geography."

8. For discussions of flagship species in the conservation literature, see Caro and Girling, *Conservation by Proxy*; Barua, "Mobilizing Metaphors."

9. The term *multispecies ethnography* was coined in Kirksey and Helmreich, "The Emergence of Multispecies Ethnography." For examples of the recent engagement with the ethology of von Uexkull, see Agamben, *The Open*; Ansell-Pearson, *Germinal Life*; Buchanan, *Onto-ethologies*; Ingold, *The Perception of the Environment*.

10. See von Uexkull and O'Neil, *A Foray onto the Worlds of Animals and Humans*.

11. Hayles, "Constrained Constructivism."

12. The etymology of *jizz* is contested. A popular attribution links the word to the corruption of an acronym borrowed from World War II aircraft spotters that refers to the general impression of shape and size (GISS) of a

plane. For discussion, see MacDonald, "What Makes You a Scientist Is the Way You Look at Things". For a more extensive discussion of the role of jizz, in natural history, see Ellis, "Jizz and the Joy of Pattern Recognition."

13. Sukumar, *The Living Elephants*; Blake and Hedges, *Sinking the Flagship.*

14. See Harrington, *Reenchanted Science.*

15. Buchanan, *Onto-ethologies.*

16. Deleuze and Guattari, *A Thousand Plateaus.*

17. Despret, "The Body We Care For"; Haraway, *When Species Meet.* For further examples, see Latour, "How to Talk about the Body."

18. See for example Lingis, *Dangerous Emotions*; Kohn, "How Dogs Dream."

19. The sociologists of science John Law and Michael Lynch discuss the role of the field guide in relation to their shared enthusiasms for bird watching. See Law and Lynch, "Pictures, Texts, and Objects."

20. There is a wide literature on the role of technologies and skilled bodies in the multispecies ethnographies of field science. See for example Helmreich, *Alien Ocean*; Candea, "I Fell in Love with Carlos the Meerkat"; Fuentes, "Naturalcultural Encounters in Bali"; Hinchliffe et al., "Urban Wild Things."

21. For influential accounts of affect, see Thrift, *Non-representational Theory*; Stewart, *Ordinary Affects*; Brian Massumi, *Parables for the Virtual.* A flavor of this field is given by Gregg and Seigworth, *The Affect Theory Reader.*

22. Lorenz, "Die Angeborenen Formen Möglicher Erfahrung." See also Serpell, "Anthropomorphism and Anthropomorphic Selection"; Gould, "A Biological Homage to Mickey Mouse."

23. Darwin, Ekman, and Prodger, *The Expression of the Emotions in Man and Animals.*

24. Kellert and Wilson, *The Biophilia Hypothesis*; Wilson, *Biophilia;* Tuan, *Topophilia.*

25. Jones, "(Un)ethical Geographies"; Wolfe, *Zoontologies.*

26. Baker, "Sloughing the Human"; Tyler, "If Horses Had Hands."

27. Acampora, *Corporal Compassion*; Bekoff, *Minding Animals.*

28. Milton, *Loving Nature.*

29. Weinstein, *Insects in Psychiatry;* Lemelin, *The Management of Insects in Recreation and Tourism*; Raffles, *Insectopedia.*

30. Hillman, *Going Bugs.*

31. Ibid., 59.

32. Deleuze and Guattari, *A Thousand Plateaus.*

33. Jones, "(Un)ethical Geographies."

34. For a representative sample of work in this genre, see the films screened at the annual Insect Fear Film Festival hosted by the Entomology Gradu-

ate Student Association at the University of Illinois at Urbana–Champaign. See the Entomology Graduate Student Association website, www.life.illinois.edu/entomology/egsa/ifff.html.

35. Hillman, "The *Satya* Interview."

36. Hillman, "Going Bugs," 60.

37. See Douglas, *Purity and Danger.*

38. Kristeva, *Powers of Horror.*

39. A surprising number of entomologists I have spoken to over the past ten years have explained to me how their interest in their target organism originally stemmed from a childhood phobia. This is especially the case for arachnologists, whose phobia inspires the ability to spot spiderlike forms, which gives them a head start when searching for arachnids in the wild.

40. Hölldobler and Wilson, *The Ants*; Goulson, *A Sting in the Tale.*

41. Baker, *The Postmodern Animal.* In *When Species Meet,* Donna Haraway detects this fear in the writings of Deleuze and Guattari. See Lorimer, "Ladies and Gentlemen."

42. Cocker, *Birders,* 19.

43. Bennett, *The Enchantment of Modern Life.* See also Fullagar, "Desiring Nature."

44. Kristeva, *Powers of Horror.* In its literal translation, *jouissance* has a more explicitly sexual or orgasmic meaning and is used as such by Henri Lefebvre. I am grateful to Eric Swyngedouw for this observation.

45. Cocker, *Birders,* 147.

46. Birding (especially in the United Kingdom) has recently been the subject of a popular publishing boom comprising a range of autobiographical accounts, how-to guides, and cultural histories (of both the taxa and individual species). Together, these give a rich flavor of the norms, values, and affective logics of this group. In addition to Mark Cocker's *Birders,* see Dee, *The Running Sky*; Birkhead, *The Wisdom of Birds*; McGrath, *Bearded Tit*; Barnes, *How to Be a Bad Birdwatcher.*

47. For a compelling articulation of this mode of intellectual satisfaction, see Gould, *Wonderful Life.*

48. For an extensive discussion and defense of the affective dimensions of abstraction and abstractions, see McCormack, "Geography and Abstraction."

3. BIODIVERSITY AS BIOPOLITICS

1. *Biodiversity* first appeared in print in 1988 as the title of the proceedings of the 1986 National Forum on Biological Diversity. See Wilson, *Biodiversity.* For an extensive genealogy of the early years of biodiversity and conservation biology, see Takacs, *The Idea of Biodiversity.*

2. Soule, *What Is Conservation Biology.*

3. HMSO, *Biodiversity.*

4. I take this phrase from Braun, "Environmental Issues".

5. See Article 2 of the Convention of Biological Diversity use of terms at http://www.cbd.int.

6. See for example Wilson, *The Diversity of Life.*

7. For a range of examples, see Shukin, *Animal Capital*; Holloway et al., "Biopower, Genetics and Livestock Breeding"; Hinchliffe et al., "Biosecurity and the Topologies of Infected Life"; Srinivasan, "The Biopolitics of Animal Being and Welfare"; Franklin, *Dolly Mixtures*; Braverman, *Zooland*; Helmreich, *Alien Ocean.*

8. Hennessy, "Producing 'Prehistoric' Life"; Buller, "Safe from the Wolf"; Yusoff, "Biopolitical Economies and the Political Aesthetics of Climate Change"; Collard, "Cougar–Human Entanglements"; Barker, "Flexible Boundaries in Biosecurity"; Clark, "Mobile Life"; Youatt, "Counting Species."

9. Mol, *Ontological Politics*; Thompson, *Making Parents.*

10. I take the metaphor of the cut from Barad, "Posthumanist Performativity."

11. This chapter draws primarily on my PhD research, conducted between 2001 and 2005. This involved a series of interviews with senior scientists and policy makers in UK conservation, textual analysis of key policy documents, and a statistical analysis of data relating to the taxonomic and geographic scope of UK conservation. It also involved three case studies of different UK species and their action plans: the corncrake, the stag beetle, and the black redstart.

12. For background information on the history and practice of British conservation, see Marren, *Nature Conservation*; Adams, *Future Nature*; Evans, *A History of Nature Conservation in Britain.*

13. See Adams, *Against Extinction*; Brockington, *Fortress Conservation*; Mackenzie, *The Empire of Nature.*

14. See Fudge, *Animal*; Dardenne and Mesplede, *A Nation of Animal Lovers?*

15. Cahill, *Who Owns Britain.* One oft-quoted statistic from this book is that 69 percent of land in the United Kingdom is in the hands of 0.6 percent of the population.

16. HMSO, *Biodiversity,* 11.

17. HMSO, *Sustaining the Variety of Life,* 9.

18. See HMSO, *Biodiversity.*

19. See the IUCN Red List website at www.iucnredlist.org.

20. Burnett et al., *Biological Recording in the United Kingdom.*

21. HMSO, *Biodiversity.*

22. Latour et al., *Paris Ville Invisible*; Bowker, "Biodiversity Datadiversity."

23. Butchart et al., *Global Biodiversity.*

24. Waterton et al., "Barcoding Nature."

25. Heywood, *Global Biodiversity Assessment,* 113.

26. Haila, "Making the Biodiversity Crisis Tractable"; Primack, *A Primer of Conservation Biology*; Gillson et al., *Base-lines, Patterns and Process.*

27. Systematics is the discipline responsible for describing species. It incorporates the identification of species, the naming and classification of species (taxonomy), and the description of the relationships among and between different taxa (phylogenetics).

28. For a fuller discussion, see Rojas, *The Species Problem and Conservation*; Wilson, *The Diversity of Life.*

29. Primack, *A Primer of Conservation Biology,* 10.

30. Wilson, *Naturalist,* 156.

31. Hunter, *Fundamentals of Conservation Biology,* 34.

32. Atran, *Cognitive Foundations of Natural History.*

33. At least this has been the popular perception for the past couple of decades. See Godfray, "Challenges for Taxonomy"; Gaston and May, "Taxonomy of Taxonomists." It has recently been the subject of some dispute and discussion. See Joppa et al., "The Population Ecology and Social Behaviour of Taxonomists."

34. Hawksworth, *Magnitude and Distribution of Biodiversity,* 116.

35. May, "Tropical Arthropod Species, More or Less?"

36. Uncertainty over the number of described species is exacerbated by the lack of any central register and the existence of high levels of synonymy in the dataset. Stork estimates that there may be some 1.8 to 2.0 million names in use by taxonomists to describe 1.2 million species. Stork, *Measuring Global Biodiversity and Its Decline.*

37. Mora et al., "How Many Species Are There on Earth and in the Ocean?"

38. Wilson, *Naturalist.*

39. See for example E. O. Wilson's 2007 TED talk, which energized the endeavor to create the *Encyclopedia of Life,* http://eol.org.

40. Gaston and May, "Taxonomy of Taxonomists," 281.

41. Mora et al., "How Many Species Are There on Earth and in the Ocean?"

42. Darwin, *The Life and Letters of Charles Darwin.*

43. HMSO, *Biodiversity: The UK Action Plan.* A new butterfly was discovered in 2001. See Nelson et al., "*Leptidea reali* Reissinger 1989." A plant was discovered in 2003. See Lowe, "A New British Species." Climate change

and anthropogenic movement also result in the arrival and settlement of new species.

44. Burnett et al., *Biological Recording in the United Kingdom*.

45. Ibid.

46. In fact, the authors of the report explain that this bird dominance was underestimated in their survey as the extensive data collections of the RSPB and the Wetlands and Wildlife Trust were not made available.

47. A good example would be the RSPB's Big Garden Birdwatch, which regularly attracts nearly 600,000 participants (close to 10 percent of the UK population).

48. Stubbs, *British Hoverflies*.

49. Bowker, "Biodiversity Datadiversity."

50. See the National Biodiversity Network website at www.nbn.org.uk.

51. For more information on these sustainable development indicators, the rationale behind their selection, and their monitoring, see www .sustainable-development.gov.uk.

52. The notable exception to this state of affairs is Butterfly Conservation, who are lead partner for all butterfly and some moth SAPs. Butterfly Conservation has recently experienced a substantial growth in membership and resources. For more information on Butterfly Conservation, see their website at http://www.butterfly-conservation.org.

53. For more information on the history, funding, and conservation work of Buglife, see their website at http://www.buglife.org.uk.

54. Bowker, "Biodiversity Datadiversity."

4. CONSERVATION AS COMPOSITION

1. Crofting is a land tenure system specific to the north of Scotland. A croft is a small holding of about five hectares, normally on a large estate. Traditionally, a crofter is a tenant farmer who pays rent to the owner of the estate. An increasing number of crofters have purchased the title to their crofts and have become owner-occupier farmers. For more information, see the Scottish Crofters Federation website at www.crofting.org.

2. This chapter draws primarily on ethnographic fieldwork conducted in the summer of 2003. During this period I spent several months in the Hebrides, speaking to crofters and conservationists, witnessing the corncrake census, and observing corncrake-friendly farming. I also conducted a series of interviews with scientists and policy makers involved with corncrake conservation outside the Hebrides.

3. I take the term *composition* from an influential article by James Baird

Callicott in which he differentiates compositionalist and functionalist approaches to conservation. The conservation I describe here maps closely on to what he describes as compositionalist approaches. See Callicott et al., "Current Normative Concepts in Conservation." A political ecology of composition has recently been promoted by Bruno Latour in his recent compositionist manifesto. Latour is rather sketchy about the ecology that would inform his "ecologized politics," but in its enthusiasm for form and human action it is well aligned with Callicott's understanding. Although the manifesto is dedicated to DH (presumably Donna Haraway), ontologically it is not that well aligned with her processural and thoroughly ecological understanding of companionship. See Haraway, *When Species Meet*.

4. The rationalization of British ecology and nature conservation is discussed in more detail in Adams, *Rationalization and Conservation*; Livingstone, "The Polity of Nature"; Morris and Reed, "From Burgers to Biodiversity?"

5. For example, the corncrake features in the nature poetry of John Clare (1793–1864). See "The Landrail" at http://www.poemhunter.com/best-poems/john-clare/the-landrail. Mrs. Beeton offers a recipe for corncrake in her famous cookbook, *Mrs. Beeton's Book of Household Management*. For a full British cultural history, see Cocker and Mabey, *Birds Britannica*.

6. Linnaeus, *Systema Naturae*.

7. Bechstein, *Ornithologisches Taschenbuch von und für Deutschland*.

8. See Alexander, "A Report on the Land-rail Enquiry"; Bannerman and Lodge, *The Birds of the British Isles*; Norris, "Report on the Corncrake"; Norris, "Report on the Distribution and Status of the Corncrake."

9. Latour, *Science in Action*.

10. Green et al., "A Simulation Model of the Effect of Mowing."

11. This summary draws on the following papers published by the RSPB scientists working on corncrakes during this period: Green, "The Decline of the Corncrake *Crex crex* in Britain Continues"; Hudson et al., "Status and Distribution of Corncrakes in Britain in 1988"; Stowe and Hudson, "Corncrake Studies in the Western Isles"; Stowe and Hudson, "Radio-Telemetry Studies of Corncrake in Great Britain"; Tyler and Green, "The Incidence of Nocturnal Song by Male Corncrakes (*Crex crex*) is Reduced during Pairing."

12. Gilbert et al., *Bird Monitoring Methods*.

13. This is the phrase used by Hudson in "Status and Distribution of Corncrakes in Britain in 1988."

14. Lynch and Law, "Pictures, Texts, and Objects"; Waterton, "From Field to Fantasy."

15. See Hinchliffe et al., "Urban Wild Things," 648.

16. For a detailed account of the epistemic properties of the field, see Kohler, *Landscapes & Labscapes*; Gieryn, "City as Truth-Spot."

17. Kohler, "Place and Practice in Field Biology."

18. Latour, *Pandora's Hope*.

19. Roth and Bowen, "Digitizing Lizards."

20. Green and Gibbons, "The Status of the Corncrake *Crex crex* in Britain in 1998."

21. Norris, "Report on the Corncrake"; Norris, "Report on the Distribution and Status of the Corncrake."

22. Kohler, *Landscapes & Labscapes*.

23. Gieryn, "City as Truth-Spot"; Guggenheim, "Laboratizing and De-laboratizing the World."

24. Guggenheim, "Laboratizing and De-laboratizing the World."

25. See Gieryn, "City as Truth-Spot."

26. Hudson, "Status and Distribution of Corncrakes in Britain in 1988"; Green et al., *A Simulation Model*; Tyler et al., "Survival and Behaviour of Corncrake."

27. In practice many of these corncrakes weren't killed but instead were escorted to safety by the watching researcher, who then had to decide whether they would have made it on their own. Fields were searched after mowing for evidence of dead birds. The effect of being observed by RSPB researchers on the corncrake-avoidance abilities of the mowing crofters is not discussed in any of the published accounts.

28. Green et al., "A Simulation Model," 106.

29. Ibid., 106.

30. Tyler et al., "Survival and Behaviour of Corncrake."

31. Green et al., "Halting Declines in Globally Threatened Species."

32. Gregory et al., *The Population Status of Birds in the United Kingdom*.

33. Subsequent monitoring in Russia found that predicted global declines in the corncrake have not taken place and the population is much more abundant and stable than previously thought. In 2004 the IUCN changed the status of the corncrake to "near threatened." In 2010 this was further downgraded to "least concern." See the entry for *Crex crex* on IUCN website at www.iucnredlist.org.

34. See the entry for *Crex crex* on IUCN website at www.iucnredlist.org.

35. Kohler, *Landscapes & Labscapes*; Henke, "Cultivating Science."

36. RSPB, *Farms, Crofts & Corncrakes*; Green and Riley, *Corncrakes*.

37. See Hunter, *The Claim of Crofting*; H. Lorimer, "Guns, Game and the Grandee."

38. RSPB, *The Future of ESAs in Scotland*.

39. Hinchliffe, "Reconstituting Nature Conservation."

40. UKBP, *Corncrake Species Action Plan.*

41. The corncrake reintroduction project is a collaboration between the RSPB and the Zoological Society of London. The corncrakes are bred at Whipsnade Zoo and released in the RSPB's reserve in the Nene Washes. It began in 2000 and now has a returning population of birds.

42. Green, "Survival and Dispersal of Male Corncrakes," 228.

43. Monbiot, *Feral;* Fraser, *Rewilding the World.*

44. Morris and Reed, "From Burgers to Biodiversity?"

45. Ritzer, *The McDonaldization Thesis.*

46. Clark and Murdoch, "Local Knowledge and the Precarious Extension of Scientific Networks."

47. Vera, *Grazing Ecology and Forest History.*

48. Harris et al., *Ecological Restoration and Global Climate Change.*

49. See Marris, *Rambunctious Garden.*

50. I take this phrase from Hinchliffe, "Reconstituting Nature Conservation."

5. WILD EXPERIMENTS

1. Vera, *Grazing Ecology and Forest History.*

2. Belt, "Networking Nature".

3. See for example the work of the NGO Rewilding Europe, which is looking to expand Vera's model of naturalistic grazing into the abandoned former agricultural landscapes of eastern Europe. See the Rewilding Europe website at www.rewildingeurope.com.

4. Marris, *Rambunctious Garden;* Balmford, *Wild Hope;* Sutherland, "Conservation Biology."

5. Navarro and Pereira, "Rewilding Abandoned Landscapes in Europe."

6. The European Commission's Eurostat website explains that "the EU budgetary spending on agro-environmental measures has increased rapidly since 1993 and it reached EUR 3 026 million in 2010. The total public funding was considerably higher (EUR 5 053 million) as Member States pay up to 50 % of the cost of measures from their own national budgets." See http://epp.eurostat.ec.europa.eu.

7. Baerselman and Vera, *Nature Development;* Belt, "Networking Nature."

8. Birks, "Mind the Gap"; Svenning, "A Review of Natural Vegetation Openness in North-western Europe."

9. See ICMO2, *Natural Processes, Animal Welfare, Moral Aspects and Management of the Oostvaardersplassen;* ICMO, *Reconciling Nature and Human Interests.*

10. Fraser, *Rewilding the World*; Monbiot, *Feral*; Donlan et al., "Pleistocene Rewilding"; Zimov, "Pleistocene Park."

11. For an overview see Balmford et al., "What Conservationists Need to Know about Farming"; Phalan et al., "Reconciling Food Production and Biodiversity Conservation." Enthusiasts take heart from the "forest transition" that is claimed to be taking place in many parts of the developed world. Here, urbanization, the abandonment of farming in marginal areas, and agricultural intensification elsewhere are resulting in significant forest grow back. See Mather, "The Forest Transition"; Rudel et al., "Agricultural Intensification and Changes in Cultivated Areas." Detractors note that on a global scale such gains are exceeded by deforestation in the developing world. See Geist and Lambin, "Proximate Causes and Underlying Driving Forces of Tropical Deforestation."

12. Gieryn, "City as Truth-Spot"; Guggenheim, "Laboratizing and Delaboratizing the World."

13. Beck, *World at Risk*; Krohn and Weyer, "Society as a Laboratory."

14. Latour, "An Attempt at a 'Compositionist Manifesto'"; Gross, *Inventing Nature.*

15. Kohler, *Landscapes & Labscapes.*

16. Rheinberger, *Toward a History of Epistemic Things.*

17. Gross, "The Public Proceduralization of Contingency."

18. Hinchliffe, "Reconstituting Nature Conservation"; Hinchliffe et al., "Urban Wild Things."

19. Callon et al., *Acting in an Uncertain World.*

20. Ibid., 178.

21. This work echoes recent writing by Latour, whose multinatural political ecology is centered on forms of cautious "public" or "collective" experimentation. See Latour, "From Multiculturalism to Multinaturalism"; Latour and Weibel, *Making Things Public*; Whatmore and Braun, *Political Matter.*

22. See Vera et al., *Wilderness in Europe.*

23. See for example Hodder et al., *Large Herbivores in the Wildwood.*

24. Vera, *Grazing Ecology and Forest History*, 24.

25. Marris, "Ecology"; Marris, "Conservation Biology."

26. Wild Europe, *Wild Europe Field Programme.*

27. Rewilding Europe, *Rewilding as a Tool.*

28. See Drenthen, "Developing Nature along Dutch Rivers."

29. See Bijker, "American and Dutch Coastal Engineering."

30. See Baerselman and Vera, *Nature Development.*

31. Latour, "An Attempt at a 'Compositionist Manifesto.'"

32. Cattle and horses are seen as the low-hanging fruit for reintroduction projects, as they are plentiful and pose little risk to property or human

health. Elsewhere, efforts have focused on predators—such as wolves in Yellowstone—or the use of surrogate species that will perform ecological functions previously performed by now extinct antecedents—for example, with tortoises in the Galapagos. More speculative projects propose the translocation and even reincarnation of keystone proboscidean species—for example, elephants (in place of sheep) in the United Kingdom or to control invasive species in Australia. There has also recently been a great deal of discussion about re-creating and returning the mammoth to Siberia.

33. There is a fascinating backstory to the Heck cattle deployed at OVP, which I explore elsewhere. In short, they were back-bred by two zoologists— Lutz and Heinz Heck—in Germany in the 1930s. Lutz in particular was a close associate of Hermann Goering, who became a patron of his project. Goering and Heck planned to reintroduce the aurochs into the future wilds of the Third Reich. See Lorimer and Driessen, "Bovine Biopolitics"; Lorimer and Driessen, "Back-breeding the Aurochs."

34. Aarden, "Natuurbeheer Dode dieren in de Oostvaardersplassen."

35. Rechtbank's-Gravenhage, "Uitspraak in Kort Geding over de Noodzaak tot Bijvoeren van Grote Grazers (o.a. Edelherten) in de Oostvaardersplassen in het Licht van Wettelijke Zorgplichten."

36. Klaver et al., "Born to Be Wild," 14. Jozef Keulartz provides a wider discussion of this process in relation to the practices of ecological restoration. See Keulartz, "The Emergence of Enlightened Anthropocentrism."

37. I take this phrase from Philo and Wilbert, *Animal Spaces, Beastly Places.*

38. Vera, *Grazing Ecology and Forest History,* 376.

39. In his writing on rationalization in British nature conservation, Bill Adams notes the challenge nonequilibrium ecology poses to the existing scientific and bureaucratic infrastructure for delivering conservation and ongoing efforts to "rerationalize" conservation in-line with the insights of this new paradigm. See Adams, "Rationalization and Conservation"; Botkin, *Discordant Harmonies.*

40. This chapter draws on research I conducted with Clemens Driessen in 2010–11. It involved interviews with those responsible for governing OVP alongside textual analysis of policy documentation, legislation, and media coverage.

41. Vulink, *Hungry Herds.*

42. See Ripple and Beschta, *Wolves and the Ecology of Fear.* The reintroduction of the wolf (or even dedomesticated dogs) and their important "ecology of fear" is seen as a step too far by the managers at OVP, but wolves have recently returned to the Netherlands under their own steam.

43. Steve Hinchliffe and his coauthors develop this notion of an epistemic

"wild thing" from Rheinberger's work on epistemic things. They apply the concept in their work on the conservation of urban water voles.

44. William Sutherland is a professor of conservation biology at Cambridge University and an influential figure in European and international conservation. For his comments on OVP, see Sutherland, *Openness in Conservation*.

45. For more information on Natura 2000, see http://ec.europa.eu.

46. ICMO, *Reconciling Nature and Human Interests*, 13.

47. Hodder et al., *Large Herbivores in the Wildwood*.

48. See Adams, "Rationalization and Conservation"; Monbiot, *Feral*.

49. Bruinderink et al., "Robuuste verbindingen en wilde hoefdieren."

50. Turnhout, "Ecological Indicators in Dutch Nature Conservation."

51. Staatsbosbeheer, *Managementplan Oostvaardersplassengebied 2011–2015*.

52. Hinchliffe et al., "Urban Wild Things"; Hinchliffe and Lavau, "Differentiated Circuits."

53. ICMO2, *Natural Processes, Animal Welfare, Moral Aspects and Management of the Oostvaardersplassen*, 84.

54. Popper, *The Open Society and Its Enemies*.

55. See www.oostvaardersplassen-sterfte.nl.

56. Smit, *Oostvaardersplassen*.

57. The film is called *The New Wilderness*. See http://www.denieuwewildernis.nl. The foxes on film are at www.volgdevos.nl.

58. Bijker, *The Oosterschelde Storm Surge Barrier*.

59. Callon et al., *Acting in an Uncertain World*.

60. Evans, "Resilience, Ecology and Adaptation in the Experimental City."

6. WILDLIFE ON SCREEN

1. Lippit, *Electric Animal*; Berger, *About Looking*.

2. Rutherford, *Governing the Wild*; Brockington, *Celebrity and the Environment*; Tsing, *Friction*.

3. See for example the charity Wildscreen (www.wildscreen.org.uk) and a critique in Brockington, *Celebrity and the Environment*.

4. For a range of examples, see Bousé, *Wildlife Films*; Mitman, *Reel Nature*; Ryan, *Picturing Empire*; Igoe et al., "A Spectacular Eco-Tour around the Historic Bloc"; Beardsworth and Bryman, "The Wild Animal in Late Modernity."

5. Burt, *Animals in Film*.

6. Lippit, *Electric Animal*. I am especially drawn to the use of Lippit's

phrase in a recent piece by Rosemary Collard that is forthcoming in the journal *Area*.

7. Miller, *Biodiversity Conservation and the Extinction of Experience*; Louv, *Last Child in the Woods*.

8. Latham and McCormack, "Thinking with Images in Non-representational Cities," 253.

9. Totaro, "Deleuzian Film Analysis."

10. The term *haptic visuality* comes from Marks, *Touch*. See also Sobchack, *Carnal Thoughts*; Bruno, *Atlas of Emotion*.

11. Haraway, *When Species Meet*, 258. Haraway borrows this phrase from her PhD student Eva Hayward. See Hayward, "Fingeryeyes."

12. Deleuze, *Cinema 1*; Deleuze, *Cinema 2*.

13. Latham and McCormack, "Thinking with Images."

14. I take these terms from Doel and Clarke, "Afterimages." See also Hansen, *Bodies in Code*.

15. Yusoff, "Biopolitical Economies and the Political Aesthetics of Climate Change," 28.

16. Rancière, *The Politics of Aesthetics,* in Yusoff, "Biopolitical Economies," 79.

17. See for example Burt, *Morbidity and Vitalism*; Cater and McCormack, "Film, Geopolitics"; Connolly, *Neuropolitics*.

18. See Anderson, "Affect and Biopower"; Thrift, "Intensities of Feeling."

19. Carter and Dodds, "Hollywood and the 'War on Terror'"; Carter and McCormack, "Film, Geopolitics and the Affective Logics of Intervention"; Shapiro, *Cinematic Geopolitics*.

20. Connolly, *Neuropolitics*, 75; Massumi, *Parables for the Virtual*.

21. Connolly, *Film Technique and Micropolitics*.

22. Connolly, *Capitalism and Christianity, American Style*.

23. Bennett and Shapiro, *The Politics of Moralizing*, 6.

24. Gibson-Graham, *A Postcapitalist Politics*; Amin and Thrift, *Arts of the Political*.

25. This is not a comprehensive survey of the different ways in which elephants are evoked—indeed, there are some notable absences, not least moving images of elephants as a source of horror and fear that could be produced by those on the receiving end of human–elephant conflict. Unfortunately, I was unable to source these.

26. The methodology for this analysis owes a great deal to existing approaches that explore images as representations (e.g., Rose, *Visual Methodologies*; Pink, *Doing Visual Ethnography*). It involves both autobiography and a reflexive awareness of the cultural and political landscape in that any

encounter with moving imagery occurs. Learning to be affected by moving imagery on a personal or group level is both straightforward and incredibly complex. It means going with the film, turning down the academic's instinct to detach, and being swept through the emotional landscape on offer. To understand how particular affects are achieved requires a constant deconstructive attention to the syntax of moving imagery—attending to types of shot, sequencing, sound, music, etc. Furthermore, to be able to speak for the wider evocative power of an image, we must situate it with the cultural norms of the audience—paying attention to both the evocation of universals and the constant possibility of confusion, transgression, or offense at an unfamiliar or unexpected response.

27. Deleuze, *Cinema 1*.

28. In his short history of envisioning elephants, Nigel Rothfels identifies the persistent focus on elephant eyes and explains how they have been used to evoke a pathetic expression of sentiment. He traces this sentimental affective logic toward elephants back to the French naturalist Buffon, writing in the mid-eighteenth century. See Rothfels, "The Eyes of Elephants."

29. For a range of essays exploring the impact and cultural politics of Disney's films, see Wilson, *The Culture of Nature*; Bell et al., *From Mouse to Mermaid*.

30. Beardsworth and Bryman argue: "In the Disneyized zoo context, the anthropomorphic and 'sentimental' expectations brought into the setting by the visitor may then be catered for by the presentation of animal performances in which creatures are invited to exhibit apparently human motivations, attributes and actions. An example is the Shamu show in Sea World theme parks in which killer whales are trained to produce behaviors which mimic human actions ('waving' at the audience, exhibiting 'shame' and 'playing' by soaking the first ten rows of onlookers)." Beardsworth and Bryman, "The Wild Animal in Late Modernity," 99.

31. Compare for example Deleuze and Guattari, *A Thousand Plateaus*; Baudrillard, *Simulacra and Simulation*; Midgley, *Animals and Why They Matter*; Scruton, *Animal Rights and Wrongs*.

32. Deleuze and Guattari, *A Thousand Plateaus*.

33. Clutton-Brock, *A Natural History of Domesticated Mammals*; Leach, *Human Domestication Reconsidered*.

34. The neologism *petishism* suggests that pets are fetishes, desired commodified objects whose allure masks the economic relations of their production. This is not an argument about animal exploitation but a critique of sentimental pet love and its perceived social, ecological, and economic consequences. Anne Friedberg coined this phrase in 1989. See Nast, "Critical Pet Studies?," for a polemical account of "pet love" and Donna Haraway, *When*

Species Meet, for a more nuanced, personal, and sympathetic description of pet keeping and its political economy.

35. Deleuze, *Cinema 1.*

36. These trippy images resonate with scenes from both *Fantasia,* Disney's avant-garde experiment, which was produced a year before *Dumbo* to little commercial success, and *Destino,* his (recently completed) collaboration with Salvador Dali.

37. Martyn Colbeck's elephant films include *Echo of the Elephants* (London, BBC, 1993); *Echo of the Elephants: The Next Generation* (London, BBC, 1996); *Echo of the Elephants: The Final Chapter?* (London, BBC, 2005); *Eye for an Elephant* (London, BBC, 2006).

38. I take the term *pachyderm personalities* from a commentary by Greg Mitman on these films. See Mitman, *Pachyderm Personalities.*

39. *Elephant Diaries* has had two series and was screened in 2005 and 2008. It featured the presenters Jonathan Scott and Michaela Strachan. The series followed the successful model established in the eight series of *Big Cat Diaries. Orangutan Diary* followed in 2007 and 2009.

40. See PETA's website at www.petatv.com.

41. *Elephants in Circuses: Training and Tragedy.* See PETA's website at www.petatv.com.

42. Hayward, "Fingeryeyes."

43. The BBC series *Spy in the Jungle* provides another excellent elephant-related example of the power of technology to create transformative media encounters. Here, captive elephants carried cameras to film tigers in National Parks.

44. Collard, "Electric Elephant."

45. Mitman, *Reel Nature.*

46. Mitman, "Pachyderm Personalities."

47. "Mammoth Journey," episode 6 of *Walking with Beasts* (London, BBC, 2001).

48. See for example a series of recent films that explore and question the narrative of heroic individuals taking on the wilderness—e.g., *Grizzly Man, Into the Wild, Brokeback Mountain.*

49. Clark, *Inhuman Nature.*

50. See a recent entry on the blog *Thinking Like a Human* by Chris Sandbrook and Bill Adams entitled "The BBC's Africa as Middle Earth." See www.thinkinglikeahuman.wordpress.com and the three-part *Unnatural Histories* series, which aired on BBC4 in 2011. The acknowledgment of the politics of framing Africa (and other places) as wilderness in these series makes it all the more surprising and galling that the big budget "blue-chip" outputs (like *Africa*) continue to ignore the histories and geographies of these

regions. This visualization of wilderness is also a technical feat. During a recent ethnography, I undertook the Wildscreen Festival in Bristol, the annual celebration of wildlife filmmaking (www.wildscreen.org.uk); one postproduction expert detailed with an insider's mixture of pride and embarrassment at the extensive manipulations required. Visual and sonic evidence of hot-air balloons, Land Rovers, pylons, and other human paraphernalia common in the National Parks where filming occurs must be removed. A soon-to-be-released Disney film on cheetahs required so much editing that he estimated at least 80 percent contained some CGI.

51. See for example Igoe et al., "A Spectacular Eco-Tour around the Historic Bloc"; Igoe, "The Spectacle of Nature."

52. These films were collected and released in 2007 as *Science Is Fiction / The Sounds of Science: The Films of Jean Painlevé* (BFI, London).

53. See for example Claude Nurisdany and Marie Perennou, *Microcosmos* (Paris, Pathe, 1996).

54. Knox, "Sounding the Depths"; Painlevé et al., *Science Is Fiction.*

55. Hayward, "Enfolded Vision."

56. Baker, *The Postmodern Animal*; Baker, *Artist Animal.*

57. Deleuze, *Cinema 2.*

58. Bill Viola, *I Do Not Know What It Is I Am Like* (Huntingdon, Quantum Leap, 1986).

59. *The Elephant in the Room* (2006 Barely Legal show, Los Angeles).

60. *Letter on the Blind, For the Use of Those Who See* (2011 Arthouse at the Jones Center, Austin Texas).

61. *The Sultan's Elephant* was a performance created by the Royal de Luxe theater company. It was commissioned to commemorate the centenary of Jules Verne's death by the two French cities of Nantes and Amiens and was performed at various locations around the world between 2005 and 2006. It involved a large moving mechanical elephant, a giant marionette of a young girl, and other public art installations.

62. Steve Hinchliffe provides a useful differentiation between care and curiosity in his appeal for a careful political ecology. See Hinchliffe, "Reconstituting Nature Conservation."

63. Bennett, *The Enchantment of Modern Life.*

64. See Reekie, *Subversion*; Russell, *Experimental Ethnography.*

7. BRINGING WILDLIFE TO MARKET

1. Elephant Parade is a separate NGO to Elephant Family. They organize outdoor exhibitions of fiberglass elephants in cities around the world. For

more information on these organizations, see www.elephantparadelondon
.org and www.elephantfamily.org.

2. Maan Barua explores the cosmopolitan character of these elephants
in a recent paper. The analysis I offer in this chapter is indebted to Maan's
work and our conversations about elephants over the past few years. See
Barua, "Circulating Elephants."

3. Haraway, *When Species Meet*. See also Rajan, *Lively Capital*.

4. Sullivan, "Banking Nature?"; Collard and Dempsey, "Life for Sale?";
Collard, "Putting Animals Back Together, Taking Commodities Apart";
Cooper, *Life as Surplus*.

5. Buscher et al., "Towards a Synthesized Critique of Neoliberal Biodiversity Conservation"; Igoe and Brockington, *Neoliberal Conservation*.

6. Brockington and Scholfield, "The Conservationist Mode of Production."

7. Adams, *Against Extinction*; Brockington and Duffy, *Capitalism and
Conservation*; Duffy, *Nature Crime*.

8. Brockington et al., *Nature Unbound*, 194. See also Igoe, "The Spectacle of Nature."

9. Neves, "Cashing in on Cetourism."

10. Tsing, *Friction*.

11. Pine and Gilmore, *The Experience Economy*. For a discussion in the
context of the commodification of Nature Inc., see Paterson, *Consumption
and Everyday Life*.

12. The concept of emotional labor was first described in Hochschild, *The
Managed Heart*. For a discussion of its role in human–animal encounters, see
Beardsworth and Bryman, "The Wild Animal in Late Modernity"; Davis,
Spectacular Nature.

13. Davis, *Spectacular Nature*; West and Carrier, "Ecotourism and Authenticity"; Bulbeck, *Facing the Wild*.

14. Carrier and Macleod, "Bursting the Bubble".

15. Rutherford, *Governing the Wild*.

16. See Beardsworth and Bryman, "The Wild Animal in Late Modernity." This claim is hotly contested by a range of critics. For a nuanced discussion of this debate, see Braverman, *Zooland*.

17. I take this information from reporting by the BBC. See www.bbc.
co.uk. For more on "panda politics," see Nicholls, *The Way of the Panda*;
Buckingham et al., "Diplomats and Refugees."

18. See "Pandas Save Edinburgh Zoo from Extinction," *Guardian*, May
9, 2013, www.guardian.co.uk.

19. See Peter Savodnik, "Knut, the $140 Million Polar Bear," www
.petersavodnik.com.

20. Zoochoses are mental and behavioral problems experienced by captive animals. The phrase was coined by Bill Travers in his 1992 documentary *The Zoochotic Report*. Travers was an actor and one of the founders of the NGO Born Free. For further discussion of these problems, see Clubb and Mason, *Captivity Effects on Wide-ranging Carnivores*; Mason and Rushen, *Stereotypic Animal Behaviour*.

21. For a detailed, Foucauldian analysis of the pastoral care associated with the modern zoo, see Braverman, *Zooland*. These techniques are also discussed in Bulbeck, *Facing the Wild*.

22. Collard, "Putting Animals Back Together."

23. Brooks et al., *Global Biodiversity Conservation Priorities*.

24. Rodriguez et al., "Environment—Globalization of Conservation."

25. "Volunteer" is something of a misnomer here, as most will pay for the placement in addition to paying for food and accommodation and providing unpaid labor. These are fee-paying volunteers. For an introduction to the sector, see Wearing, *Volunteer Tourism*.

26. Bröckerhoff and Wadham-Smith, *Volunteering: Global Citizenship in Action*.

27. This argument is put forward by sending organizations and has been picked up in a series of reports commissioned by the UK government. See for example Jones, *Review of Gap Year Provision*.

28. Honey, *Ecotourism and Sustainable Development*.

29. Duffy, *A Trip Too Far*; Simpson, "Doing Development"; Tomazos and Butler, "The Volunteer Tourist as Hero."

30. One notable exception is the work of Jennifer Cousins. See Cousins et al., "I've Paid to Observe Lions, Not Map Roads!"

31. This involved a survey of the main organizations sending UK volunteers overseas, interviews with their staff, and a detailed case study of volunteers involved in Asian elephant conservation in Sri Lanka. It offers a snapshot of the sector. I have published the findings of this survey in more detail elsewhere. See Lorimer, "International Volunteering from the UK."

32. See Lorimer, "International Conservation 'Volunteering' and the Geographies of Global Environmental Citizenship."

33. For a detailed discussion of the history of elephant hunting and safari in Sri Lanka, see Lorimer and Whatmore, "After 'The King of Beasts.'"

34. I discuss this relationship in more detail in Lorimer, "Touching Environmentalisms."

35. For a review of this field, see Stone, "Dark Tourism Scholarship."

36. Weber, *The Sociology of Charismatic Authority*. See also Weber and Eisenstadt, *Max Weber on Charisma and Institution Building*.

37. Star and Griesemer, "Institutional Ecology."

38. Jalais, *Unmasking the Cosmopolitan Tiger.*

39. Barua, *The Political Ecology of Human–Elephant Relationships in India.*

40. Menon, *Right of Passage.*

41. Barua, *The Political Ecology of Human–Elephant Relationships in India,* 17.

42. Jalais, *Forest of Tigers.*

43. I take the mercury metaphor from Clark, "Money Flows Like Mercury."

44. Barua, *The Political Ecology of Human–Elephant Relationships in India.*

45. I take this phrase from Robertson, "The Nature That Capital Can See."

46. Collard and Dempsey, "Life for Sale?"

47. The concept of a virtualism describes a world performed out of the abstract logics of market rationality. For an introduction to the concept, see Carrier and Miller, *Virtualism.* For its application to conservation, see Brockington and Duffy, *Capitalism and Conservation;* Carrier and West, *Virtualism, Governance and Practice;* Brockington, Duffy, and Igoe, *Nature Unbound.*

8. SPACES FOR WILDLIFE

1. For more information about Living Roofs, see www.livingroofs.org.

2. I take the term *unofficial countryside* from Richard Mabey's influential book about urban wildlife in the United Kingdom. See Mabey, *The Unofficial Countryside.*

3. I take the term *edgelands* from Michael Symmons Roberts and Paul Farley, *Edgelands.*

4. For different accounts of this recent interest in urban natures, see Whatmore, "Living Cities"; Heynen et al., *In the Nature of Cities;* Gaston, *Urban Ecology.*

5. The World Database on Protected Areas gives a wealth of information on the scope and character of protected areas. For analysis and discussion, see Soutullo, "Extent of the Global Network of Terrestrial Protected Areas"; Jenkins and Joppa, "Expansion of the Global Terrestrial Protected Area System." The coverage of marine areas is much less extensive (about 1 percent). See Toropova et al., "Global Ocean Protection."

6. For an account of this history, see Adams, *Against Extinction.*

7. For a range of examples of the history of protected areas in different national contexts, see Mackenzie, *The Empire of Nature;* Jazeel, "'Nature,' Nationhood and the Poetics"; Jepson and Whittaker, "Histories of Protected Areas"; Bhagwat and Rutte, "Sacred Groves."

8. See MacArthur and Wilson, *The Theory of Island Biogeography.*

9. For accounts of the influence of island biogeography on the designation and management of conservation territories, see Takacs, *The Idea of Biodiversity*; Whittaker et al., "Conservation Biogeography."

10. Brockington, *Fortress Conservation.*

11. These arguments are made in different ways by Cronon, "The Trouble with Wilderness"; Botkin, *Discordant Harmonies*; Katz, "Whose Nature, Whose Culture?"; Whatmore and Thorne, "Wild(er)ness."

12. Wolch, "Zoopolis."

13. Hinchliffe, "Cities and Natures."

14. Hinchliffe et al., "Urban Wild Things," 645.

15. Mol and Law, Regions, "Networks and Fluids," 647.

16. I take the concept of the territorial trap from Agnew, "The Territorial Trap."

17. For more information about the specificities of urban ecologies and novel ecosystems, see Francis and Chadwick, "The Ecology of Urban Environments"; Francis, Lorimer, and Raco, "Urban Ecosystems."

18. For a discussion of the concept of synurbic species, see Francis and Chadwick, "What Makes a Species Synurbic?"

19. See for example Zalasiewicz et al., *The Anthropocene.*

20. I take the term *everyday environmentalism* from Loftus, *Everyday Environmentalism.*

21. I understand the term *green urbanism* to describe an influential, technocratic approach to urban planning that emerged in European cities in the late 1990s that sought to forge "sustainable cities." For an overview see Beatley, *Green Urbanism.*

22. See for example reports on this campaign on Buglife's website at www.buglife.org.uk.

23. This document provides an exemplary appeal for green urbanism. See Urban Task Force, *Towards an Urban Renaissance.*

24. For an illuminating discussion of recent developments in the aesthetics and ecology of postindustrial urban wastelands, see Gandy, "Marginalia."

25. See discussions in Gandy, "Queer Ecology"; Sinclair, *Ghost Milk*; Roberts and Farley, *Edgelands.*

26. Chipchase and Frith, *Brownfield? Greenfield?*

27. Hitchings, "How Awkward Encounters Could Influence the Future Form of Many Gardens."

28. Hengeveld, *Biodiversity*; Key, *Bare Ground and the Conservation of Invertebrates.*

29. Law and Mol, "Situating Technoscience," 614.

30. A note on terminology. The term *living roofs* is used by practitioners

to describe both "intensive" green roofs with grass, soil, and trees and "extensive" roofs created from rubble and favoring xerophytic organisms. The latter are normally brown. Dusty explained that he often calls the roofs "green" even when they are brown to emphasize their environmentally beneficial or "green" credentials. In terms of wildlife, brown roofs are often "greener" than their green counterparts, though they may have other benefits for urban hydrology, climatology, and amenity.

31. EcoSchemes, *Green Roofs*; Gedge and Kadas, *Green Roofs and Biodiversity*; Francis and Lorimer, "Urban Reconciliation Ecology."

32. Steve Hinchliffe gives a more extended discussion of this speculative approach to urban wildlife conservation in his appeals for a careful political ecology. Drawing on similar case studies and contemporaneous research to that documented in this chapter, he traces how figures like Dusty persuaded planners and developers to accommodate the "likely presence" of key species in the creation of new urban habitats. See Hinchliffe, "Reconstituting Nature Conservation."

33. In September 2013 the British government published a consultation paper on biodiversity offsetting, alongside a range of documents explaining how such schemes would be implemented. See www.gov.uk/biodiversity-offsetting. For a critical review of these developments, see Hannis and Sullivan, "Offsetting Nature?"; Sullivan, "Banking Nature?"

34. Law and Mol, "Situating Technoscience," 615.

35. See for example a creative agency afforded fire in Fuhlendorf et al., "Pyric Herbivory."

36. See for example the Lawton review that informed the recent UK Government White Paper on the Natural Environment. Lawton, *Making Space for Nature*. Further enthusiasm for connectivity can be seen in the creation of the Connectivity Conservation Thematic Group at the IUCN and the proliferation of textbooks, conferences, and special editions of conservation journals.

37. Rosenzweig, *Win-Win Ecology*.

38. Parmesan and Yohe, "A Globally Coherent Fingerprint of Climate Change Impacts Across Natural Systems"; Mawdsley et al., "A Review of Climate-Change Adaptation Strategies for Wildlife Management and Biodiversity Conservation."

39. See for example Whatmore and Thorne, "Elephants on the Move"; Murdoch, *Post-structuralist Geography*; Smith, "World City Actor-Networks."

40. Ingold, *Being Alive*.

41. Deleuze and Guattari, *A Thousand Plateaus*.

42. See for example landscape efforts by British NGOs in the interests

of hedgehogs (www.hedgehogstreet.org); B-Lines, for bees, butterflies, and other invertebrates (www.buglife.org.uk); and ponds for frogs and other aquatic life (www.pondconservation.org.uk).

43. Baerselman and Vera, *Nature Development*; Belt, "Networking Nature."

44. See Warren, "Perspectives on the 'Alien' versus 'Native.'"

45. Hinchliffe et al., "Biosecurity and the Topologies of Infected Life"; Dobson et al., "Biosecurity."

46. Mason, *Connecting Canals*.

47. See also Barker, "Flexible Boundaries in Biosecurity."

48. Hinchliffe, "Reconstituting Nature Conservation"; Hinchliffe et al., "Biosecurity and the Topologies of Infected life"; Hinchliffe et al., "Urban Wild Things"; Hinchliffe and Lavau, "Differentiated Circuits."

49. Ingold, *Lines*, 103.

CONCLUSION

1. I take the concept of anticipatory semantics for the Anthropocene from a lecture given by Noel Castree in Cardiff in 2013.

2. I take the concept of flourishing from Haraway, who develops it from the work of ecofeminist philosophers like Chris Cuomo and Val Plumwood. Cuomo modifies Aristotle's transcendent, humanist model of flourishing to propose a more-than-human account that values the immanent tendencies and affective force—or what she terms the "dynamic charm" (Cuomo, *Feminism and Ecological Communities*, 71) of individual nonhumans and the aggregates they compose. Haraway reframes dynamic charm as a sense of "response-ability," which describes both an ability to adapt to resist change and the ways in which such adaptations draw others into a relationship. See Haraway, *When Species Meet*; Cuomo, *Feminism and Ecological Communities*; Plumwood, *Environmental Culture*.

3. Haraway, "When Species Meet: Staying with the Trouble."

4. Marris, *Rambunctious Garden*.

5. There is a diverse environmental literature that takes the garden as a metaphor for exploring human–environment relationships. For a range of examples, see Pollan, *Second Nature*; Howard, *Garden Cities of Tomorrow*; Marx, *The Machine in the Garden*.

6. Haraway, *When Species Meet*, 17.

7. IUCN, *IUCN Red List of Threatened Species*.

8. Fleischer et al., "Phylogeography of the Asian Elephant."

9. Fernando et al., "DNA Analysis Indicates That Asian Elephants Are Native to Borneo."

10. For a comparable of these "complexities" in African elephant conser-

vation, see Thompson, "When Elephants Stand for Competing Philosophies of Nature."

11. Monbiot, *Feral;* Donlan et al., "Pleistocene Rewilding."

12. Zimov, "Pleistocene Park."

13. Bowman, "Conservation: Bring Elephants to Australia?"

14. For a review, see Caro, "The Pleistocene Re-wilding Gambit."

15. Haraway, *When Species Meet*; Yusoff, "Aesthetics of Loss."

16. For an extended discussion of these (and other modes) of posthumanism, see Lorimer, "Posthumanism/Posthumanistic Geographies."

17. Braun, "Towards a New Earth and a New Humanity," 219.

18. Braun, "Environmental Issues."

19. See for example Hinchliffe et al., "Biosecurity and the Topologies of Infected Life"; Hinchliffe et al., "Urban Wild Things."

20. For a discussion of such future-oriented modes of management, see Hinchliffe, "Reconstituting Nature Conservation"; Braun, "Biopolitics and the Molecularization of Life"; Anderson, "Preemption, Precaution, Preparedness."

21. Latour, "An Attempt at a 'Compositionist Manifesto'"; Latour, "Love Your Monsters."

22. The environmental historian Donald Worster notes the political convenience of a fluid ecology of nonequilibrium for capitalist destruction, in the same volume that eminent conservation biologist Michael Soule asserts that some forms of poststructuralist philosophy are as ecologically damaging as chainsaws. See Soule and Lease, *Reinventing Nature?* James Evans notes the ontological similarities between fungible, laissez-faire, neoliberal natures and fluid, self-willed ecologies. See Evans, "Resilience, Ecology and Adaptation in the Experimental City".

23. For an introduction and critical overview, see Sullivan, "Banking Nature?"

24. See Cooper, *Life as Surplus*; Walker and Cooper, "Genealogies of Resilience."

25. See for example Robertson, "Measurement and Alienation"; McAfee, "Selling Nature to Save It?"; Brockington and Duffy, *Capitalism and Conservation*; Hayden, *When Nature Goes Public.*

26. Robertson, "The Nature That Capital Can See."

27. Bowker, "Time, Money and Biodiversity."

BIBLIOGRAPHY

Acampora, Ralph R. *Corporal Compassion: Animal Ethics and Philosophy of Body*. Pittsburgh: University of Pittsburgh Press, 2006.

Adams, William. *Against Extinction: The Story of Conservation*. London: Earthscan, 2004.

———. *Future Nature: A Vision for Conservation*. London: Taylor & Francis, 2012.

———. "Rationalization and Conservation: Ecology and the Management of Nature in the United Kingdom." *Transactions of the Institute of British Geographers* 22, no. 3 (1997): 277–91.

Agamben, Giorgio. *The Open: Man and Animal*. Stanford, Calif.: Stanford University Press, 2004.

Agnew, John. "The Territorial Trap: The Geographical Assumptions of International Relations Theory." *Review of International Political Economy* 1 (1994): 53–80.

Agrawal, Arun. *Environmentality: Technologies of Government and the Making of Subjects*. Durham, N.C.: Duke University Press, 2005.

Alexander, H. "A Report on the Land-Rail Enquiry." *British Birds* 8 (1914): 82–92.

Alexander, K. A., E. Pleydell, M. C. Williams, E. P. Lane, J. F. C. Nyange, and A. L. Michel. "Mycobacterium Tuberculosis: An Emerging Disease of Free-Ranging Wildlife." *Emerging Infectious Diseases* 8 (2002): 598–601.

Amin, Ash, and Nigel Thrift. *Arts of the Political: New Openings for the Left*. Durham, N.C.: Duke University Press, 2013.

Anderson, Ben. "Affect and Biopower: Towards a Politics of Life." *Transactions of the Institute of British Geographers* 37 (2012): 28–43.

———. "Preemption, Precaution, Preparedness: Anticipatory Action and Future Geographies." *Progress in Human Geography* 34, no. 6 (December 1, 2010): 777–98.

Anderson, Ben, Matthew Kearnes, Colin McFarlane, and Dan Swanton. "On Assemblages and Geography." *Dialogues in Human Geography* 2, no. 2 (2012): 171–89.

Ansell-Pearson, Keith. *Germinal Life: The Difference and Repetition of Deleuze.* New York: Routledge, 1999.

Atran, Scott. *Cognitive Foundations of Natural History: Towards an Anthropology of Science.* Cambridge: Cambridge University Press, 1990.

Avise, John C. *Phylogeography: The History and Formation of Species.* Cambridge: Harvard University Press, 2000.

Baerselman, F., and F. Vera. *Nature Development: An Exploratory Study for the Construction of Ecological Networks.* The Hague, Netherlands: Ministry of Agriculture, Nature Management, and Fisheries, 1995.

Baker, S. *Artist Animal.* Minneapolis: University of Minnesota Press, 2013.

———. *The Postmodern Animal.* London: Reaktion Books, 2000.

———. "Sloughing the Human." In *Zoontologies: The Question of the Animal,* edited by C. Woolf, 147–64. Minneapolis: University of Minnesota Press, 2003.

Balmford, Andrew. *Wild Hope: On the Front Lines of Conservation Success.* Chicago: University of Chicago Press, 2012.

Balmford, Andrew, Rhys Green, and Ben Phalan. "What Conservationists Need to Know about Farming." *Proceedings of the Royal Society B: Biological Sciences* 279, no. 1739 (2012): 2714–24.

Bandara, R., and C. Tisdell. "Comparison of Rural and Urban Attitudes to the Conservation of Asian Elephants in Sri Lanka: Empirical Evidence." *Biological Conservation* 110, no. 3 (2003): 327–42.

Bannerman, David Armitage, and George Edward Lodge. *The Birds of the British Isles.* Edinburgh: Oliver and Boyd, 1953.

Barad, Karen. "Posthumanist Performativity: Toward an Understanding of How Matter Comes to Matter." *Signs: Journal of Women in Culture and Society* 28, no. 3 (2003).

Barker, Kezia. "Flexible Boundaries in Biosecurity: Accomodating Gorse in Aotearoa New Zealand." *Environment and Planning A* 40, no. 7 (2008): 1598–614.

Barnes, Simon. *How to Be a Bad Birdwatcher.* London: Short Books, 2005.

Barua, Maan. "Circulating Elephants: Unpacking the Geographies of a Cosmopolitan Animal." *Transactions of the Institute of British Geographers* 39, no. 4 (2013): 559–73.

———. "Mobilizing Metaphors: The Popular Use of Keystone, Flagship and Umbrella Species Concepts." *Biodiversity and Conservation* 20, no. 7 (2011): 1427–40.

———. "The Political Ecology of Human-Elephant Relationships in India." PhD diss., University of Oxford, 2013.

Bates, L. A., K. N. Sayialel, N. W. Njiraini, C. J. Moss, J. H. Poole, and

R. W. Byrne. "Elephants Classify Human Ethnic Groups by Odor and Garment Color." *Current Biology* 17, no. 22 (2007): 1938–42.

Baudrillard, Jean. *Simulacra and Simulation.* Ann Arbor: University of Michigan Press, 1994.

Bear, Chris, and Sally Eden. "Thinking Like a Fish? Engaging with Nonhuman Difference through Recreational Angling." *Environment and Planning D: Society and Space* 29, no. 2 (2011): 336–52.

Beardsworth, Alan, and Alan Bryman. "The Wild Animal in Late Modernity: The Case of the Disneyization of Zoos." *Tourist Studies* 1, no. 1 (2001): 83–104.

Beatley, Timothy. *Green Urbanism: Learning from European Cities.* Washington, D.C.: Island Press, 1999.

Bechstein, Johann. *Ornithologisches Taschenbuch von und für Deutschland oder kurze Beschreibung aller Vogel Deutschlands.* Vol. 2. Leipzig: Richter, 1803.

Beck, Ulrich. *World at Risk.* Cambridge: Polity, 2009.

Bekoff, Marc. *Minding Animals: Awareness, Emotions, and Heart.* New York: Oxford University Press, 2002.

Bell, Elizabeth, Lynda Haas, and Laura Sells. *From Mouse to Mermaid: The Politics of Film, Gender, and Culture.* Bloomington: Indiana University Press, 1995.

Belt, Henk van den. "Networking Nature, or Serengeti behind the Dikes." *History and Technology* 20, no. 3 (2004): 311–33.

Bengtsson, J., P. Angelstam, T. Elmqvist, U. Emanuelsson, C. Folke, M. Ihse, F. Moberg, and M. Nystram. "Reserves, Resilience and Dynamic Landscapes." *Ambio* 32, no. 6 (2003): 389–96.

Bennett, Jane. "The Agency of Assemblages and the North American Blackout." *Public Culture* 17, no. 3 (2005): 445–66.

———. *The Enchantment of Modern Life: Attachments, Crossings, and Ethics.* Princeton, N.J.: Princeton University Press, 2001.

———. "The Force of Things: Steps toward an Ecology of Matter." *Political Theory* 32, no. 3 (June 2004): 347–72.

———. *Vibrant Matter: A Political Ecology of Things.* Durham, N.C.: Duke University Press, 2010.

Bennett, Jane, and Michael J. Shapiro. *The Politics of Moralizing.* New York: Routledge, 2002.

Berger, John. *About Looking.* New York: Vintage International, 1980.

Bhagwat, Shonil A., and Claudia Rutte. "Sacred Groves: Potential for Biodiversity Management." *Frontiers in Ecology and the Environment* 4, no. 10 (2006): 519–24.

Bijker, Wiebe E. "American and Dutch Coastal Engineering: Differences in Risk Conception and Differences in Technological Culture." *Social Studies of Science* 37, no. 1 (2007): 143–51.

———. "The Oosterschelde Storm Surge Barrier: A Test Case for Dutch Water Technology, Management, and Politics." *Technology and Culture* 43 (2002): 569–84.

Birkhead, Tim. *The Wisdom of Birds: An Illustrated History of Ornithology.* London: Bloomsbury, 2008.

Birks, H. John. "Mind the Gap: How Open Were European Primeval Forests?" *Trends in Ecology and Evolution* 20, no. 4 (2005): 154–56.

Blake, Stephen, and Simon Hedges. "Sinking the Flagship: The Case of Forest Elephants in Asia and Africa." *Conservation Biology* 18, no. 5 (2004): 1191–202.

Botkin, Daniel. *Discordant Harmonies: A New Ecology for the Twenty-First Century.* New York: Oxford University Press, 1990.

Bousé, Derek. *Wildlife Films.* Philadelphia: University of Pennsylvania Press, 2000.

Bowker, Geoffrey. "Biodiversity Datadiversity." *Social Studies of Science* 30, no. 5 (2000): 643–83.

———. "Time, Money, and Biodiversity." In *Global Assemblages: Technology, Politics, and Ethics as Anthropological Problems,* edited by Aiwa Ong and Stephen Collier, 107–23. Oxford: Blackwell, 2005.

Bowman, David. "Conservation: Bring Elephants to Australia?" *Nature* 482, no. 7383 (2012): 30.

Bradshaw, Gay. *Elephants on the Edge: What Animals Teach Us about Humanity.* New Haven, Conn.: Yale University Press, 2009.

———. "Not by Bread Alone: Symbolic Loss, Trauma, and Recovery in Elephant Communities." *Society and Animals* 12, no. 2 (2004): 143–58.

Brand, Stewart. *Whole Earth Discipline: An Ecopragmatist Manifesto.* New York: Viking, 2009.

Braun, Bruce. "Biopolitics and the Molecularization of Life." *Cultural Geographies* 14, no. 1 (2007): 6–28.

———. "Environmental Issues: Global Natures in the Space of Assemblage." *Progress in Human Geography* 30, no. 5 (2006): 644–54.

———. "Environmental Issues: Inventive Life." *Progress in Human Geography* 32, no. 5 (2008): 667–79.

———. "Towards a New Earth and a New Humanity: Nature, Ontology, Politics." In *David Harvey: A Critical Reader,* edited by N. Castree and D. Gregory, 191–222. Oxford: Blackwell, 2006.

Braverman, Irus. *Zooland: The Institution of Captivity.* Stanford, Calif.: Stanford University Press, 2012.

Bright, Chris. *Life Out of Bounds: Bioinvasion in a Borderless World*. New York: Norton, 1998.

Bröckerhoff, Aurélie., and Nick Wadham-Smith. *Volunteering: Global Citizenship in Action*. London: Counterpoint, 2008.

Brockington, Dan. *Celebrity and the Environment: Fame, Wealth and Power in Conservation*. London: Zed Books, 2009.

———. *Fortress Conservation: The Preservation of the Mkomazi Game Reserve, Tanzania*. Bloomington: Indiana University Press, 2002.

Brockington, Daniel, and Rosaleen Duffy. *Capitalism and Conservation*. Malden, Mass.: Wiley-Blackwell, 2011.

Brockington, Dan, Rosaleen Duffy, and Jim Igoe. *Nature Unbound: Conservation, Capitalism and the Future of Protected Areas*. London: Earthscan, 2008.

Brockington, Dan, and Katherine Scholfield. "The Conservationist Mode of Production and Conservation Ngos in Sub-Saharan Africa." *Antipode* 42, no. 3 (2010): 551–75.

Brooks, T. M., R. A. Mittermeier, G. A. B. da Fonseca, J. Gerlach, M. Hoffmann, J. F. Lamoreux, C. G. Mittermeier, J. D. Pilgrim, and A. S. L. Rodrigues. "Global Biodiversity Conservation Priorities." *Science* 313, no. 5783 (2006): 58–61.

Bruno, Giuliana. *Atlas of Emotion: Journeys in Art, Architecture, and Film*. London: Verso Books, 2002.

Buchanan, Brett. *Onto-ethologies: The Animal Environments of Uexküll, Heidegger, Merleau-Ponty, and Deleuze*. Albany: SUNY Press, 2008.

Buckingham, Kathleen Carmel, Jonathan Neil William David, and Paul Jepson. "Diplomats and Refugees: Panda Diplomacy, Soft 'Cuddly' Power, and the New Trajectory in Panda Conservation." *Environmental Practice* (2013): 1–9.

Bulbeck, Chilla. *Facing the Wild: Ecotourism, Conservation, and Animal Encounters*. London: Earthscan, 2005.

Buller, Henry. "Safe from the Wolf: Biosecurity, Biodiversity, and Competing Philosophies of Nature." *Environment and Planning A* 40, no. 7 (2008): 1583–97.

Burnett, J., Copp, C. J. T., and Harding P. T. *Biological Recording in the United Kingdom: Present Practice and Future Development*. Vol. 1. London: Joint Nature Conservation Committee, 1995.

Burt, John. *Animals in Film*. London: Reaktion, 2002.

Buscher, Bram, Sian Sullivan, Katja Neves, Jim Igoe, and Dan Brockington. "Towards a Synthesized Critique of Neoliberal Biodiversity Conservation." *Capitalism Nature Socialism* 23, no. 2 (2012): 4–30.

Butchart, Stuart H. M., Matt Walpole, Ben Collen, Arco van Strien, Jörn

P. W. Scharlemann, Rosamunde E. A. Almond, Jonathan E. M. Baillie, et al. "Global Biodiversity: Indicators of Recent Declines." *Science* 328, no. 5982 (2010): 1164–68.

Cabinet Office. *The Future Role of the Third Sector in Social and Economic Regeneration: Final Report.* London: HMSO, 2007.

Cahill, Kevin. *Who Owns Britain.* Edinburgh: Canongate Books, 2002.

Callicott, J. Baird, Lary Crowder, and Karen Mumford. "Current Normative Concepts in Conservation." *Conservation Biology* 13, no. 1 (1999): 22–35.

Callon, Michel, Pierre Lascoumes, and Yannick Barthe. *Acting in an Uncertain World: An Essay on Technical Democracy.* Cambridge, Mass.: MIT Press, 2009.

Candea, Matei. "'I Fell in Love with Carlos the Meerkat': Engagement and Detachment in Human–Animal Relations." *American Ethnologist* 37, no. 2 (2010): 241–58.

Caro, Tim. *Conservation by Proxy: Indicator, Umbrella, Keystone, Flagship, and Other Surrogate Species.* Washington, D.C.: Island Press, 2010.

———. "The Pleistocene Re-wilding Gambit." *Trends in Ecology & Evolution* 22, no. 6 (2007): 281–83.

Caro, Tim, Jack Darwin, Tavis Forrester, Cynthia Ledoux-Bloom, and Caitlin Wells. "Conservation in the Anthropocene." *Conservation Biology* 26, no. 1 (2012): 185–88.

Carrier, James, and Daniel Miller. *Virtualism: A New Political Economy.* Oxford: Berg, 1998.

Carrier, James, and Paige West. *Virtualism, Governance and Practice: Vision and Execution in Environmental Conservation.* Oxford: Berg, 2009.

Carrier, James G., and Donald V. L. Macleod. "Bursting the Bubble: The Socio-cultural Context of Ecotourism." *Journal of the Royal Anthropological Institute* 11, no. 2 (2005): 315–34.

Carter, Sean, and Klaus Dodds. "Hollywood and the 'War on Terror': Genre-Geopolitics and 'Jacksonianism' in the Kingdom." *Environment and Planning D: Society and Space* 29, no. 1 (2011): 98–113.

Carter, Sean, and Derek McCormack. "Film, Geopolitics, and the Affective Logics of Intervention." *Political Geography* 25, no. 2 (2006): 228–45.

Cassidy, Rebecca, and Molly H. Mullin. *Where the Wild Things Are Now: Domestication Reconsidered.* Oxford: Berg, 2007.

Chakrabarty, Dipresh. "The Climate of History: Four Theses." *Critical Inquiry* 35, no. 2 (2009): 197–222.

Chipchase, Annie, and Matthew Frith. *Brownfield? Greenfield? The Threat to London's Unofficial Countryside.* London: London Wildlife Trust, 2002.

Clark, Gordon "Money Flows Like Mercury: The Geography of Global Finance." *Geografiska Annaler: Series B, Human Geography* 87, no. 2 (2005): 99–112.

Clark, Judy, and Jonathan Murdoch. "Local Knowledge and the Precarious Extension of Scientific Networks: A Reflection on Three Case Studies." *Sociologia Ruralis* 37, no. 1 (1997): 38–60.

Clark, Nigel. "The Demon-Seed: Bioinvasion as the Unsettling of Environmental Cosmopolitanism." *Theory Culture & Society* 19, no. 1–2 (2002): 101–25.

———. *Inhuman Nature: Sociable Living on a Dynamic Planet.* Thousand Oaks, Calif.: Sage, 2011.

———. "Mobile Life: Biosecurity Practices and Insect Globalization." *Science as Culture* 22, no. 1 (2013): 16–37.

Clark, Nigel, Arun Saldanha, and Kathryn Yusoff, eds. *Capitalism and the Earth.* Brooklyn, N.Y.: Punctum Books, 2014.

Clubb, Rosemary, and Georgina Mason. "Captivity Effects on Wide-Ranging Carnivores." *Nature* 425, no. 6957 (2003): 473–74.

Clutton-Brock, Juliet. *A Natural History of Domesticated Mammals.* Cambridge: Cambridge University Press, 1999.

Cocker, Mark. *Birders: Tales of a Tribe.* London: Jonathan Cape, 2001.

Cocker, Mark, and Richard Mabey. *Birds Britannica.* London: Chatto & Windus, 2005.

Collard, Rosemary-Caire. "Cougar–Human Entanglements and the Biopolitical Un/Making of Safe Space." *Environment and Planning D: Society and Space* 30, no. 1 (2012): 23–42.

———. "Putting Animals Back Together, Taking Commodities Apart." *Annals of the Association of American Geographers* 104, no. 1 (2014): 151–65.

Collard, Rosemary-Claire, and Jessica Dempsey. "Life for Sale? The Politics of Lively Commodities." *Environment and Planning A* 45, no. 11 (2013): 2682–99.

Connolly, William. *Capitalism and Christianity, American Style.* Durham, N.C.: Duke University Press, 2008.

———. "Film Technique and Micropolitics." *Theory and Event* 6, no. 1 (2002).

———. *Neuropolitics: Thinking, Culture, Speed.* Minneapolis: University of Minnesota Press, 2002.

Cooper, Melinda. *Life as Surplus: Biotechnology and Capitalism in the Neoliberal Era.* Seattle: University of Washington Press, 2008.

Cousins, Jessica, James Evans, and Jon Sadler. "'I've Paid to Observe Lions, Not Map Roads!' An Emotional Journey with Conservation Volunteers in South Africa." *Geoforum* 40, no. 6 (2009): 1069–80.

Cronon, William. "The Trouble with Wilderness; or, Getting Back to the Wrong Nature." In *Uncommon Ground: Rethinking the Human Place in Nature,* edited by William Cronon, 69–90. New York: Norton, 1996.

———, ed. *Uncommon Ground: Rethinking the Human Place in Nature.* New York: W. W. Norton, 1996.

Crooks, K. R., and A. Sanjayan. *Connectivity Conservation.* Cambridge: Cambridge University Press, 2006.

Crutzen, Paul. "Geology of Mankind." *Nature* 415, no. 6867 (2002): 23.

Cuomo, Chris J. *Feminism and Ecological Communities: An Ethic of Flourishing.* London: Routledge, 1998.

Daily, G. C., P. R. Ehrlich, and G. A. Sanchez-Azofeifa. "Countryside Biogeography: Use of Human-Dominated Habitats by the Avifauna of Southern Costa Rica." *Ecological Applications* 11, no. 1 (2001): 1–13.

Dalby, Simon. "Biopolitics and Climate Security in the Anthropocene." *Geoforum* 49 (2013): 184–92.

Dardenne, Emilie, and Sophie Mesplede, eds. *A Nation of Animal Lovers? Representing Human–Animal Relations in Britain.* Manchester: Manchester University Press, forthcoming.

Darier, Eric. *Discourses of the Environment.* Oxford: Blackwell, 1998.

Darwin, Charles. *The Life and Letters of Charles Darwin.* London: Echo Library, 2007.

Darwin, Charles, Paul Ekman, and Phillip Prodger. *The Expression of the Emotions in Man and Animals.* London: HarperCollins, 1999.

Davies, Gail. "Caring for the Multiple and the Multitude: Assembling Animal Welfare and Enabling Ethical Critique." *Environment and Planning D: Society and Space* 30, no. 4 (2012): 623–38.

Davis, Mark, Matthew Chew, Richard Hobbs, Ariel Lugo, John Ewel, Geerat Vermeij, James Brown, et al. "Don't Judge Species on Their Origins." *Nature* 474, no. 7350 (2011): 153–54.

Davis, Susan. *Spectacular Nature: Corporate Culture and the Sea World Experience.* Berkeley: University of California Press, 1997.

Dee, Tim. *The Running Sky: A Bird-Watching Life.* Random House, 2010.

De Landa, Manuel. *Intensive Science and Virtual Philosophy.* London: Continuum, 2002.

———. *A Thousand Years of Nonlinear History.* New York: Zone Books, 1997.

Deleuze, Gilles. *Cinema 1.* London: Continuum, 2005.

———. *Cinema 2.* London: Continuum, 2005.

———. *Difference and Repetition.* London: Athlone Press, 1994.

Deleuze, Gilles, and Félix Guattari. *A Thousand Plateaus: Capitalism and Schizophrenia.* Minneapolis: University of Minnesota Press, 1987.

Derrida, Jacques, and Marie-Louise Mallet. *The Animal That Therefore I Am.* New York: Fordham University Press, 2008.

Despret, Vincienne. "The Body We Care For: Figures of Anthropo-Zoo-Genesis." *Body & Society* 10, nos. 2–3 (2004): 111–34.

Dewsbury, J. D. "The Deleuze-Guattarian Assemblage: Plastic Habits." *Area* 43, no. 2 (2011): 148–53.

Diprose, Rosalyn. *Corporeal Generosity: On Giving with Nietzsche, Merleau-Ponty, and Levinas.* Albany: SUNY Press, 2002.

Dobson, Andrew, Kezia Barker, and Sarah Tayor. *Biosecurity: The Sociopolitics of Invasive Species and Infectious Diseases.* London: Routledge, 2013.

Doel, Marcus, and David Clarke. "Afterimages." *Environment and Planning D: Society and Space* 25, no. 5 (2007): 890–910.

Donlan, C. J., J. Berger, C. E. Bock, J. H. Bock, D. A. Burney, J. A. Estes, D. Foreman, et al. "Pleistocene Rewilding: An Optimistic Agenda for Twenty-First Century Conservation." *American Naturalist* 168, no. 5 (2006): 660–81.

Douglas, Mary. *Purity and Danger: An Analysis of Concepts of Pollution and Taboo.* London: Routledge, 1966.

Drenthen, Martin. "Developing Nature along Dutch Rivers: Place or Nonplace." In *New Visions of Nature: Complexity and Authenticity,* edited by Martin Drenthen, Jozef Keulartz, and James Proctor, 205–28. London: Springer, 2009.

Duffy, Rosaleen. "Interactive Elephants: Nature, Tourism and Neoliberalism." *Annals of Tourism Research* 44 (2014): 88–101.

———. *Nature Crime: How We're Getting Conservation Wrong.* New Haven, Conn.: Yale University Press, 2010.

———. *A Trip Too Far: Ecotourism, Politics, and Exploitation.* London: Earthscan, 2002.

Duffy, Rosaleen, and Lorraine Moore. "Neoliberalising Nature? Elephant-Back Tourism in Thailand and Botswana." *Antipode* 42, no. 3 (2010): 742–66.

EcoSchemes. "Green Roofs: Their Existing Status and Potential for Conserving Biodiversity in Urban Areas." In *English Nature Research Reports—No. 498.* Peterborough, U.K.: English Nature, 2003.

Ellis, Erle. "Anthropogenic Transformation of the Terrestrial Biosphere." *Philosophical Transactions of the Royal Society A: Mathematical, Physical and Engineering Sciences* 369, no. 1938 (2011): 1010–35.

———. "Sustaining Biodiversity and People in the World's Anthropogenic Biomes." *Current Opinion in Environmental Sustainability* 5, no. 3–4 (2013): 368–72.

Ellis, Erle, and Navin Ramankutty. "Putting People in the Map: Anthropogenic Biomes of the World." *Frontiers in Ecology and the Environment* 6, no. 8 (2008): 439–47.

Ellis, Rebecca. "Jizz and the Joy of Pattern Recognition: Virtuosity, Discipline and the Agency of Insight in UK Naturalists' Arts of Seeing." *Social Studies of Science* 41, no. 6 (2011): 769–90.

Esposito, Roberto. *Immunitas: The Protection and Negation of Life*. Cambridge: Polity, 2011.

Evans, David. *A History of Nature Conservation in Britain*. London: Routledge, 1997.

Evans, James. "Resilience, Ecology and Adaptation in the Experimental City." *Transactions of the Institute of British Geographers* 36, no. 2 (2011): 223–37.

Fernando, Prithviraj. "Elephants in Sri Lanka: Past, Present and Future." *Loris* 22, no. 2 (2000): 38–44.

Fernando, Prithviraj, and Richard Lande. "Molecular Genetic and Behavioral Analysis of Social Organization in the Asian Elephant (*Elephas maximus*)." *Behavioral Ecology and Sociobiology* 48, no. 1 (2000): 84–91.

Fernando, P., M. E. Pfrender, S. E. Encalada, and R. Lande. "Mitochondrial DNA Variation, Phylogeography and Population Structure of the Asian Elephant." *Heredity* 84, no. 3 (2000): 362–72.

Fernando, P., T. N. C. Vidya, J. Payne, M. Stuewe, G. Davison, R. J. Alfred, P. Andau, et al. "DNA Analysis Indicates That Asian Elephants Are Native to Borneo and Are Therefore a High Priority for Conservation." *PLoS Biology* 1, no. 1 (2003).

Fernando, P., E. D. Wikramanayake, H. K. Janaka, L. K. A. Jayasinghe, M. Gunawardena, S. W. Kotagama, D. Weerakoon, and J. Pastorini. "Ranging Behavior of the Asian Elephant in Sri Lanka." *Mammalian Biology* 73, no. 1 (2008): 2–13.

Fernando, P., E. D. Wikramanayake, D. Weerakoon, L. K. A. Jayasinghe, M. Gunawardene, and H. K. Janaka. "Perceptions and Patterns of Human–Elephant Conflict in Old and New Settlements in Sri Lanka: Insights for Mitigation and Management." *Biodiversity and Conservation* 14, no. 10 (2005): 2465–81.

Fleischer, R. C., E. A. Perry, K. Muralidharan, E. E. Stevens, and C. M. Wemmer. "Phylogeography of the Asian Elephant (*Elephas maximus*) Based on Mitochondrial DNA." *Evolution* 55, no. 9 (2001): 1882–92.

Foucault, Michel. *The Birth of Biopolitics: Lectures at the College De France, 1978—1979*. New York: Picador, 2010.

———. *The History of Sexuality: The Will to Knowledge*. Vol. 1. New York: Penguin, 1998.

———. *Security, Territory, Population: Lectures at the Collège De France, 1977–78.* New York: Palgrave Macmillan, 2007.

———. *Society Must Be Defended: Lectures at the Collège De France, 1975–76.* New York: Picador, 2003.

Fowler, Murray, and Susan Mikota. *Biology, Medicine, and Surgery of Elephants.* Oxford: Blackwell, 2006.

Francis, Rob, and Michael Chadwick. *The Ecology of Urban Environments.* London: Taylor & Francis, 2013.

———. "What Makes a Species Synurbic?" *Applied Geography* 32, no. 2 (2012): 514–21.

Francis, Rob, and Jamie Lorimer. "Urban Reconciliation Ecology: The Potential of Living Roofs and Walls." *Journal of Environmental Management* 92, no. 6 (2011): 1429–37.

Francis, Robert, Jamie Lorimer, and Mike Raco. "Urban Ecosystems as 'Natural' Homes for Biogeographical Boundary Crossings." *Transactions of the Institute of British Geographers* 37, no. 2 (2012): 183–90.

Franklin, Sarah. *Dolly Mixtures: The Remaking of Genealogy.* Durham, N.C.: Duke University Press, 2007.

Fraser, Caroline. *Rewilding the World: Dispatches from the Conservation Revolution.* New York: Metropolitan Books, 2009.

Fudge, Erica. *Animal.* Reaktion Books, 2002.

Fuentes, Agustin. "Naturalcultural Encounters in Bali: Monkeys, Temples, Tourists, and Ethnoprimatology." *Cultural Anthropology* 25, no. 4 (2010): 600–24.

Fuhlendorf, S. D., D. M. Engle, J. Kerby, and R. Hamilton. "Pyric Herbivory: Rewilding Landscapes through the Recoupling of Fire and Grazing." *Conservation Biology* 23, no. 3 (2009): 588–98.

Fullagar, Simone. "Desiring Nature: Identity and Becoming in Narratives of Travel." *Cultural Values* 4, no. 1 (2000): 58–76.

Game, Anne. "Riding: Embodying the Centaur." *Body & Society* 7, no. 4 (2001): 1–12.

Gandy, Matthew. "Marginalia: Aesthetics, Ecology, and Urban Wastelands." *Annals of the Association of American Geographers* 103, no. 6 (2013): 1301–16.

———. "Queer Ecology: Nature, Sexuality, and Heterotopic Alliances." *Environment and Planning D: Society and Space* 30, no. 4 (2012): 727–47.

Gaston, Kevin, and Robert May. "Taxonomy of Taxonomists." *Nature* 356, no. 6367 (1992): 281–82.

Gaston, Kevin. *Biodiversity: A Biology of Numbers and Difference.* Cambridge, Mass.: Blackwell, 1996.

———. *Urban Ecology.* Cambridge: Cambridge University Press, 2010.

Gedge, Dusty, and Gyongyver Kadas. "Green Roofs and Biodiversity." *Biologist* 52, no. 3 (2005): 164–71.

Geist, H. J., and E. F. Lambin. "Proximate Causes and Underlying Driving Forces of Tropical Deforestation." *BioScience* 52, no. 2 (2002): 143–50.

Gibson-Graham, J. K. *A Postcapitalist Politics.* Minneapolis: University of Minnesota Press, 2006.

Gieryn, Thomas F. "City as Truth-Spot." *Social Studies of Science* 36, no. 1 (2006): 5–38.

Gilbert, G., Gibbons, D., Evans, J. *Bird Monitoring Methods.* Sandy, U.K.: RSPB, 1998.

Gillson, Lindsey , Richard J. Ladle, and Miguel B. Araújo. "Base-Lines, Patterns and Process." In *Conservation Biogeography,* edited by Richard J. Ladle and Robert J. Whittaker, 31–44. London: Wiley, 2011.

Godfray, H. Charles. "Challenges for Taxonomy." *Nature* 417, no. 6884 (2002): 17–19.

Gould, Stephen J. "A Biological Homage to Mickey Mouse." In *The Panda's Thumb,* edited by S. J. Gould. New York: W. W. Norton, 1980.

———. *Wonderful Life: The Burgess Shale and the Nature of History.* New York: Random House, 2000.

Goulson, Dave. *A Sting in the Tale.* New York: Random House, 2013.

Green, R. E., G. A. Tyler, T. J. Stowe, and A. V. Newton. "A Simulation Model of the Effect of Mowing of Agricultural Grassland on the Breeding Success of the Corncrake (*Crex crex*)." *Journal of Zoology* 243 (1997): 81–115.

Green, Rhys. "The Decline of the Corncrake *Crex crex* in Britain Continues." *Bird Study* 42 (1995): 66–75.

———. "Survival and Dispersal of Male Corncrakes *Crex crex* in a Threatened Population." *Bird Study* 46 (1999): 218–29.

Green, Rhys, and David Gibbons. "The Status of the Corncrake *Crex crex* in Britain in 1998." *Bird Study* 47 (2000): 129–37.

Green, Rhys, and Helen Riley. *Corncrakes.* Naturally Scottish. Perth: Scottish Natural Heritage, 1999.

Greenhough, Beth. "Where Species Meet and Mingle: Endemic Human–Virus Relations, Embodied Communication and More-than-Human Agency at the Common Cold Unit 1946–90." *Cultural Geographies* 19, no. 3 (2012): 281–301.

Greenhough, Beth, and Emma Roe. "Ethics, Space, and Somatic Sensibilities: Comparing Relationships between Scientific Researchers and Their Human and Animal Experimental Subjects." *Environment and Planning D: Society and Space* 29, no. 1 (2011): 47–66.

Gregg, M., and G. J. Seigworth. *The Affect Theory Reader.* Durham, N.C.: Duke University Press, 2010.

Gregory, R, N. Wilkinson, D. Noble, J. Robinson, A. Brown, J. Hughes, D. Procter, D. Gibbons, and C. Galbraith. "The Population Status of Birds in the United Kingdom, Channel Islands and Isle of Man: An Analysis of Conservation Concern 2002–7." *British Birds* 95 (2002): 410–48.

Gross, Mathias. *Ignorance and Surprise: Science, Society, and Ecological Design.* Cambridge, Mass.: MIT Press, 2010.

———. *Inventing Nature: Ecological Restoration by Public Experiments.* Lanham, Md.: Lexington Books, 2003.

———. "The Public Proceduralization of Contingency: Bruno Latour and the Formation of Collective Experiments." *Social Epistemology* 24, no. 1 (2010): 63–74.

Grove, Kevin. "Insuring 'Our Common Future'? Dangerous Climate Change and the Biopolitics of Environmental Security." *Geopolitics* 15, no. 3 (2010): 536–63.

Guggenheim, Michael. "Laboratizing and De-laboratizing the World." *History of the Human Sciences* 25, no. 1 (2012): 99–118.

Haila, Yrjö. "Making the Biodiversity Crisis Tractable." In *Philosophy and Biodiversity,* edited by Markku Oksanen and Juhani Pietarinen, 54–82. Cambridge: Cambridge University Press, 2004.

Hamilton, Clive. *Earthmasters: Playing God with the Climate.* Crows Nest, Australia: Allen & Unwin, 2013.

Hannis, Mike, and Sian Sullivan. *Offsetting Nature? Habitat Banking and Biodiversity Offsets in the English Land Use Planning System.* Dorset, U.K.: Green House, 2012.

Hansen, Mark. *Bodies in Code: Interfaces with Digital Media.* London: Taylor & Francis, 2012.

Haraway, Donna. *The Companion Species Manifesto: Dogs, People, and Significant Otherness.* Chicago: Prickly Paradigm Press, 2003.

———. *Simians, Cyborgs, and Women: The Reinvention of Nature.* New York: Routledge, 1991.

———. *When Species Meet.* Posthumanities. Minneapolis: University of Minnesota Press, 2008.

———. "When Species Meet: Staying with the Trouble." *Environment and Planning D: Society and Space* 28, no. 1 (2010): 53–55.

Harrington, Anne. *Reenchanted Science: Holism in German Culture from Wilhelm II to Hitler.* Princeton, N.J.: Princeton University Press, 1999.

Hawksworth, D. L. "Magnitude and Distribution of Biodiversity." In *Global Biodiversity Assessment,* edited by V. H. Heywood, 107–92. Cambridge: Cambridge University Press, 1995.

Hayden, Cori. *When Nature Goes Public: The Making and Unmaking of Bioprospecting in Mexico.* Princeton, N.J.: Princeton University Press, 2003.

Hayles, Katherine. "Constrained Constructivism: Locating Scientific Inquiry in the Theater of Representation." In *Realism and Representation: Essays on the Problem of Realism in Relation to Science, Literature, and Culture,* edited by George Levine, 27–43. Madison: University of Wisconsin Press, 1993.

Hayward, Eva. "Enfolded Vision: Refracting the Love Life of the Octopus." *Octopus: A Journal of Visual Studies* 1, no. 1 (2005): 29–44.

———. "Fingeryeyes: Impressions of Cup Corals." *Cultural Anthropology* 25, no. 4 (2010): 577–99.

Helmreich, Stefan. *Alien Ocean: Anthropological Voyages in Microbial Seas.* Berkeley: University of California Press, 2009.

Hengeveld, R. "Biodiversity: The Diversification of Life in a Non-equilibrium World." *Biodiversity Letters* 2, no. 1 (1994): 1–10.

Henke, Christopher. *Cultivating Science, Harvesting Power: Science and Industrial Agriculture in California.* Cambridge, Mass.: MIT Press, 2008.

Hennessy, Elizabeth. "Producing 'Prehistoric' Life: Conservation Breeding and the Remaking of Wildlife Genealogies." *Geoforum* 49 (2013): 71–80.

Hewitt, N., N. Klenk, A. L. Smith, D. R. Bazely, N. Yan, S. Wood, J. I. MacLellan, C. Lipsig-Mumme, and I. Henriques. "Taking Stock of the Assisted Migration Debate." *Biological Conservation* 144, no. 11 (2011): 2560–72.

Heynen, Nik, Maria Kaika, and Eric Swyngedouw. *In the Nature of Cities: Urban Political Ecology and the Politics of Urban Metabolism.* London: Taylor & Francis, 2005.

Heywood, Vernon. *Global Biodiversity Assessment.* Cambridge: Cambridge University Press, 1995.

Higgs, Eric. *Nature by Design: People, Natural Process, and Ecological Restoration.* Cambridge, Mass.: MIT Press, 2003.

Hillman, James. "Going Bugs." *Spring: A Journal of Archetype and Culture* (1988): 40–72.

———. "The *Satya* Interview: Going Bugs with James Hillman." *Satya,* January 1997, http://www.satyamag.com/jan97/going.html.

Hinchliffe, Steve. "Cities and Natures: Intimate Strangers." In *Unsettling Cities,* edited by J. Allen et al., 137–80. London: Open University Press, 1999.

———. *Geographies of Nature: Societies, Environments, Ecologies.* London: Sage, 2007.

———. "Reconstituting Nature Conservation: Towards a Careful Political Ecology." *Geoforum* 39, no. 1 (2008): 88–97.

Hinchliffe, Steve, John Allen, Stephanie Lavau, Nick Bingham, and Simon Carter. "Biosecurity and the Topologies of Infected Life: From Borderlines to Borderlands." *Transactions of the Institute of British Geographers* 38, no. 4 (2013): 531–43.

Hinchliffe, Steve, Matthew Kearnes, Monica Degen, and Sarah Whatmore. "Urban Wild Things: A Cosmopolitical Experiment." *Environment and Planning D: Society and Space* 23, no. 5 (2005): 643–58.

Hinchliffe, Steve, and Stephanie Lavau. "Differentiated Circuits: The Ecologies of Knowing and Securing Life." *Environment and Planning D: Society and Space* 31, no. 2 (2013): 259–74.

Hinchliffe, Steve, and Sarah Whatmore. "Living Cities: Towards a Politics of Conviviality." *Science as Culture* 15, no. 2 (2006): 123–38.

Hird, Myra. *The Origins of Sociable Life: Evolution after Science Studies.* Basingstoke, U.K.: Palgrave Macmillan, 2009.

Hitchings, Russell. "How Awkward Encounters Could Influence the Future Form of Many Gardens." *Transactions of the Institute of British Geographers* 32, no. 3 (2007): 363–76.

HMSO. *Biodiversity: The UK Action Plan.* London: HM Stationery Office, 1994.

———. *Biodiversity: The UK Steering Group Report.* Vol. 1. London: HM Stationery Office, 1995.

———. *Sustaining the Variety of Life: 5 Years of the UK Biodiversity Action Plan.* London: HM Stationery Office, 2001.

Hobbs, Richard, Eric Higgs, and Carol Hall. *Novel Ecosystems: Intervening in the New Ecological World Order.* London: Wiley, 2013.

Hochschild, Arlie. *The Managed Heart: Commercialization of Human Feeling.* Berkeley: University of California Press, 1983.

Hodder, K. H., J. M. Bullock, P. C. Buckland, and K. J. Kirby. "Large Herbivores in the Wildwood and Modern Naturalistic Grazing Systems." In *English Nature Research Reports.* Peterborough, U.K.: English Nature 2005.

Hölldobler, Bert, and Edward Wilson. *The Ants.* Cambridge, Mass.: Belknap Press of Harvard University Press, 1990.

Holloway, Lewis, Carol Morris, Ben Gilna, and David Gibbs. "Biopower, Genetics and Livestock Breeding: (Re)constituting Animal Populations and Heterogeneous Biosocial Collectivities." *Transactions of the Institute of British Geographers* 34, no. 3 (2009): 394–407.

Honey, Martha. *Ecotourism and Sustainable Development: Who Owns Paradise?* Washington, D.C.: Island Press, 2008.

Howard, Ebeneezer. *Garden Cities of Tomorrow.* London: Forgotten Books, 1960.

Hudson, A., T. Stowe, and S. Aspinall. "Status and Distribution of Corncrakes in Britain in 1988." *British Birds* 83 (1990): 173–87.

Hulme, Mike. *Why We Disagree about Climate Change: Understanding Controversy, Inaction and Opportunity.* Cambridge: Cambridge University Press, 2009.

Human Microbiome Project. "Structure, Function and Diversity of the Healthy Human Microbiome." *Nature* 486, no. 7402 (2012): 207–14.

Hunter, James. *The Claim of Crofting: The Scottish Highlands and Islands, 1930–1990.* Edinburgh: Mainstream, 1991.

Hunter, Malcolm L. *Fundamentals of Conservation Biology.* Cambridge, Mass.: Blackwell Science, 1996.

ICMO. *Reconciling Nature and Human Interests: Advice of the International Committee on the Management of Large Herbivores in the Oostvaardersplassen.* The Hague, Netherlands: Wageningen, 2006.

ICMO2. *Natural Processes, Animal Welfare, Moral Aspects and Management of the Oostvaardersplassen: Report of the Second International Commission on Management of the Oostvaardersplassen.* The Hague, Netherlands: Wageningen, 2010.

Igoe, Jim. "The Spectacle of Nature in the Global Economy of Appearances: Anthropological Engagements with the Spectacular Mediations of Transnational Conservation." *Critique of Anthropology* 30, no. 4 (2010): 375–97.

Igoe, Jim, and Dan Brockington. "Neoliberal Conservation: A Brief Introduction." *Conservation and Society* 5, no. 4 (2007): 432–49.

Igoe, Jim, Katja Neves, and Dan Brockington. "A Spectacular Eco-Tour around the Historic Bloc: Theorising the Convergence of Biodiversity Conservation and Capitalist Expansion." *Antipode* 42, no. 3 (2010): 486–512.

Ingold, Tim. *Being Alive: Essays on Movement, Knowledge and Description.* London: Routledge, 2011.

———. *Lines: A Brief History.* London: Taylor & Francis, 2007.

———. *The Perception of the Environment: Essays on Livelihood, Dwelling and Skill.* London: Routledge, 2000.

IUCN. *IUCN Red List of Threatened Species: 2009.* Vol. 1. Gland, Switzerland: IUCN, 2009.

Jalais, Annu. *Forest of Tigers: People, Politics and Environment in the Sundarbans.* London: Taylor & Francis, 2011.

———. "Unmasking the Cosmopolitan Tiger." *Nature and Culture* 3, no. 1 (2008): 25–40.

Jazeel, Tariq. "'Nature,' Nationhood and the Poetics of Meaning in Ruhuna (Yala) National Park, Sri Lanka." *Cultural Geographies* 12, no. 2 (April 2005): 199–227.

Jenkins, Clinton, and Lucas Joppa. "Expansion of the Global Terrestrial

Protected Area System." *Biological Conservation* 142, no. 10 (2009): 2166–74.

Jepson, Paul, and Robert Whittaker. "Histories of Protected Areas: Internationalisation of Conservationist Values and Their Adoption in the Netherlands Indies (Indonesia)." *Environment and History* 8, no. 2 (2002): 129–72.

Jones, Andrew. "Review of Gap Year Provision." London: Department for Education and Skills, 2004.

Jones, Owain. "(Un)ethical Geographies of Human–Non-human Relations: Encounters, Collectives and Spaces." In *Animal Spaces, Beastly Places,* edited by C. Philo and C. Wilbert, 268–91. London: Routledge, 2000.

Jongman, R. H. G., M. Kulvik, and I. Kristiansen. "European Ecological Networks and Greenways." *Landscape and Urban Planning* 68, nos. 2–3 (2004): 305–19.

Joppa, Lucas, David Roberts, and Stuart Pimm. "The Population Ecology and Social Behaviour of Taxonomists." *Trends in Ecology & Evolution* 26, no. 11 (2011): 551–53.

Kareiva, Peter. "Conservation in the Anthropocene." *Breakthrough Journal* 2 (2011).

Katz, Cindi. "Whose Nature, Whose Culture? Private Productions of Space and the 'Preservation' of Nature." In *Remaking Reality: Nature at the Millennium,* edited by B. Braun and N. Castree, 46–63. London: Routledge, 1998.

Kellert, Stephen, and Edward Wilson. *The Biophilia Hypothesis.* Washington, D.C.: Island Press, 1993.

Keulartz, Jozef. "The Emergence of Enlightened Anthropocentrism in Ecological Restoration." *Nature and Culture* 7, no. 1 (2012): 48–71.

Key, Roger. "Bare Ground and the Conservation of Invertebrates." *British Wildlife* 11, no. 3 (2000).

Kirksey, S. Eben, and Stefan Helmreich. "The Emergence of Multispecies Ethnography." *Cultural Anthropology* 25, no. 4 (2010): 545–76.

Klaver, I., J. Keulartz, H. Van den Belt, and B. Gremmen. "Born to Be Wild: A Pluralistic Ethics concerning Introduced Large Herbivores in the Netherlands." *Environmental Ethics* 24, no. 1 (2002): 3–21.

Knox, Jim. "Sounding the Depths: Jean Painleve's Sunken Cinema." *Senses of Cinema* 3 (2003).

Kohler, Robert. *Landscapes & Labscapes: Exploring the Lab–Field Border in Biology.* Chicago: University of Chicago Press, 2002.

———. "Place and Practice in Field Biology." *History of Science* 40, no. 128 (2002): 189–210.

Kohn, Eduardo. "How Dogs Dream: Amazonian Natures and the Politics of Transspecies Engagement." *American Ethnologist* 34, no. 1 (2007): 3–24.

Kristeva, Julia. *Powers of Horror: An Essay on Abjection.* European Perspectives. New York: Columbia University Press, 1982.

Krohn, Wolfgang, and Johannes Weyer. "Society as a Laboratory: The Social Risks of Experimental Research." *Science and Public Policy* 21, no. 3 (1994): 173–83.

Laland, Kevin N., and Bennett G. Galef. *The Question of Animal Culture.* Cambridge, Mass.: Harvard University Press, 2009.

Lane, S. N., N. Odoni, C. Landstrom, S. J. Whatmore, N. Ward, and S. Bradley. "Doing Flood Risk Science Differently: An Experiment in Radical Scientific Method." *Transactions of the Institute of British Geographers* 36, no. 1 (2011): 15–36.

Latham, Alan, and Derek McCormack. "Thinking with Images in Non-representational Cities: Vignettes from Berlin." *Area* 41, no. 3 (2009): 252–62.

Latour, Bruno. "An Attempt at a 'Compositionist Manifesto.'" *New Literary History* 41, no. 3 (2010): 471–90.

———. "How to Talk about the Body? The Normative Dimension of Science Studies." *Body and Society* 10, no. 2 (2004): 205–29.

———. "Love Your Monsters." In *Love Your Monsters: Postenvironmentalism and the Anthropocene,* edited by Michael Shellenberger and Ted Nordhaus. Oakland, Calif.: Breakthrough Institute, 2011.

———. "To Modernise or Ecologise? That Is the Question." In *Remaking Reality: Nature at the Millennium,* edited by Bruce Braun and Noel Castree, 221–42. London: Routledge, 1998.

———. "From Multiculturalism to Multinaturalism: What Rules of Method for the New Socio-scientific Experiments?" *Nature + Culture* 6, no. 1 (Spring 2011): 1–17.

———. *Pandora's Hope: Essays on the Reality of Science Studies.* Cambridge, Mass.: Harvard University Press 1999.

———. *Politics of Nature: How to Bring the Sciences into Democracy.* Cambridge, Mass.: Harvard University Press, 2004.

———. *Reassembling the Social: An Introduction to Actor-Network-Theory.* Oxford: Oxford University Press, 2005.

———. *Science in Action: How to Follow Scientists and Engineers through Society.* Maidenhead, U.K.: Open University Press, 1987.

———. *We Have Never Been Modern.* Cambridge, Mass.: Harvard University Press, 1993.

———. "Why Has Critique Run out of Steam? From Matters of Fact to Matters of Concern." *Critical Inquiry* 30, no. 2 (2004): 225–48.

Latour, Bruno, Emilie Hermant, and Susanna Shannon. *Paris Ville Invisible.* Paris: La Dâecouverte, 1998.

Latour, Bruno, and Peter Weibel. *Making Things Public: Atmospheres of Democracy.* Cambridge, Mass.: MIT Press, 2005.

Law, John, and Anne-Marie Mol. "Situating Technoscience: An Inquiry into Spatialities." *Environment and Planning D: Society and Space* 19, no. 5 (2001): 609–21.

Lawrence, Anthony, and Graham Spence. *The Elephant Whisperer.* London: Macmillan, 2009.

Lawton, John. *Making Space for Nature: A Review of England's Wildlife Sites and Ecological Network.* London: DEFRA, 2010.

Leach, Melissa. "Human Domestication Reconsidered." *Current Anthropology* 44, no. 3 (2003): 349–68.

Lemelin, Raynald. *The Management of Insects in Recreation and Tourism.* Cambridge: Cambridge University Press, 2012.

Lemke, Thomas. *Biopolitics: An Advanced Introduction.* New York: New York University Press, 2011.

Lien, Marianne Elisabeth, and John Law. "'Emergent Aliens': On Salmon, Nature, and Their Enactment." *Ethnos* 76, no. 1 (2011): 65–87.

Lingis, Alphonso. *Dangerous Emotions.* Berkeley: University of California Press, 2000.

Linnaeus, Carolus. *Systema Naturae.* 10th ed. 1758.

Lippit, Akira Mizuta. *Electric Animal: Toward a Rhetoric of Wildlife.* Minneapolis: University of Minnesota Press, 2000.

Livingstone, David. "The Polity of Nature: Representation, Virtue, Strategy." *Cultural Geographies* 2, no. 4 (1995): 353–77.

Loftus, Alex. *Everyday Environmentalism: Creating an Urban Political Ecology.* Minneapolis: University of Minnesota Press, 2012.

Lorenz, Konrad. "Die Angeborenen Formen MöGlicher Erfahrung" (The Innate Form of Possible Experience). *Zeitschrift für Tierpsychologie* 5 (1943): 235–409.

Lorimer, Hayden. "Cultural Geography: The Busyness of Being 'More-than-Representational.'" *Progress in Human Geography* 29, no. 1 (January 2005): 83–94.

———. "Guns, Game and the Grandee: The Cultural Politics of Deerstalking in the Scottish Highlands." *Ecumene* 7, no. 4 (2000): 403–31.

———. "Herding Memories of Humans and Animals." *Environment and Planning D: Society and Space* 24, no. 4 (2006): 497–518.

Lorimer, Jamie. "Elephants as Companion Species: The Lively Biogeographies of Asian Elephant Conservation in Sri Lanka." *Transactions of the Institute of British Geographers* 35, no. 4 (2010): 491–506.

———. "International Conservation 'Volunteering' and the Geographies of Global Environmental Citizenship." *Political Geography* 29, no. 6 (2010): 311–22.

———. "International Volunteering from the UK: What Does It Contribute?" *Oryx* 43, no. 3 (2009): 1–9.

———. "Ladies and Gentlemen, Behold the Enemy!" *Environment and Planning D: Society and Space* 28, no. 1 (2010): 40–42.

———. "Living Roofs and Brownfield Wildlife: Towards a Fluid Biogeography of UK Nature Conservation." *Environment and Planning A* 40, no. 9 (2008): 2042–60.

———. "Posthumanism/Posthumanistic Geographies." *International Encyclopedia of Human Geography* 8 (2009): 344–54.

———. "Touching Environmentalisms: The Place of Touch in the Fraught Biogeographies of Elephant Care." In *Touching Space, Placing Touch,* edited by M. Dodge and M. Patterson, 169–90. Farnham, U.K.: Ashgate, 2012.

Lorimer, Jamie, and Clemens Driessen. "Back-Breeding the Aurochs: The Heck Brothers, National Socialism and Imagined Geographies for Nonhuman Lebensraum." In *Hitler's Geographies: The Spatialities of the Third Reich,* edited by P. Giaccaria and C. Minca. Chicago: University of Chicago Press, forthcoming.

———. "Bovine Biopolitics and the Promise of Monsters in the Rewilding of Heck Cattle." *Geoforum* 48 (2013): 249–59.

Lorimer, Jamie, and Sarah Whatmore. "After 'The King of Beasts': Samuel Baker and the Embodied Historical Geographies of His Elephant Hunting in Mid-19th Century Ceylon." *Journal of Historical Geography* 35 (2009): 668–89.

Louv, Richard. *Last Child in the Woods: Saving Our Children from Nature-Deficit Disorder.* Chapel Hill, N.C.: Algonquin Books, 2008.

Lowe, A., and R. Abbott. "A New British Species, *Senecio eboracensis* (Asteraceae), Another Hybrid Derivative of *S. vulgaris L.* and *S. squalidus L.*" *Watsonia* 24 (2003): 1–13.

Lowe, Celia. *Wild Profusion: Biodiversity Conservation in an Indonesian Archipelago.* Princeton, N.J.: Princeton University Press, 2006.

Lynas, Mark. *The God Species: How the Planet Can Survive the Age of Humans.* London: Fourth Estate, 2011.

Lynch, Michael, and John Law. "Pictures, Texts, and Objects: The Literary Language Game of Birdwatching." In *Routledge Science Studies Reader,* edited by Mario Biagioli, 317–41. London: Routledge, 1999.

Mabey, Richard. *The Unofficial Countryside.* London: Collins, 1973.

MacArthur, Robert, and Edward Wilson. *The Theory of Island Biogeography.* Princeton, N.J.: Princeton University Press, 1967.

MacDonald, Heather. "'What Makes You a Scientist Is the Way You Look at Things': Ornithology and the Observer 1930–1955." *Studies in History and Philosophy of Biological and Biomedical Sciences* 33 (2002): 53–77.

Mackenzie, John. *The Empire of Nature: Hunting, Conservation, and British Imperialism.* Manchester: Manchester University Press, 1997.

Manning, A. D., J. Fischer, A. Felton, B. Newell, W. Steffen, and D. B. Lindenmayer. "Landscape Fluidity: A Unifying Perspective for Understanding and Adapting to Global Change." *Journal of Biogeography* 36, no. 2 (2009): 193–99.

Margulis, Lynn, and Dorion Sagan. *Acquiring Genomes: A Theory of the Origins of Species.* New York: Basic Books, 2002.

Marks, Laura. *Touch: Sensuous Theory and Multisensory Media.* Minneapolis: University of Minnesota Press, 2002.

Marren, Peter. *Nature Conservation.* London: Harper Collins, 2002.

Marris, Emma. "Conservation Biology: Reflecting the Past." *Nature* 462, no. 7269 (2009): 30–32.

———. "Ecology: Ragamuffin Earth." *Nature* 460, no. 7254 (2009): 450–53.

———. *Rambunctious Garden: Saving Nature in a Post-wild World.* New York: Bloomsbury, 2011.

Martin, Lauren, and Anna J. Secor. "Towards a Post-mathematical Topology." *Progress in Human Geography* 38, no. 3 (2014): 420–38.

Marx, Karl. *Capital: Volume 1.* London: Penguin, 1976.

Marx, Leo. *The Machine in the Garden: Technology and the Pastoral Ideal in America.* New York: Oxford University Press, 2000.

Mason, Georgina, and Jeffry Rushen. *Stereotypic Animal Behaviour: Fundamentals and Applications to Welfare.* Wallingford, U.K.: CABI, 2008.

Mason, Vicky. "Connecting Canals: Exercises in Recombinant Ecology." PhD diss., Oxford University, 2013.

Massey, Doreen. *For Space.* London: Sage, 2005.

Massumi, Brian. *Parables for the Virtual: Movement, Affect, Sensation.* Durham, N.C.: Duke University Press, 2002.

Mather, Alexander. "The Forest Transition." *Area* 24, no. 4 (1992): 367–79.

Mawdsley, Jonathan, Robin O'Malley, and Dennis Ojima. "A Review of Climate-Change Adaptation Strategies for Wildlife Management and Biodiversity Conservation." *Conservation Biology* 23, no. 5 (2009): 1080–89.

May, Robert. "Tropical Arthropod Species, More or Less?" *Science* 329, no. 5987 (2010): 41–42.

McCormack, Derek. "Geography and Abstraction: Towards an Affirmative Critique." *Progress in Human Geography* 36, no. 6 (2012): 715–34.

McGrath, Rory. *Bearded Tit.* London: Ebury Press, 2009.

McKibben, Bill. *The End of Nature.* New York: Anchor Books, 1990.

Menon, Vivek, Sandeep Kumar Tiwari, P. S. Easa, and R. Sukumar, eds. *Right of Passage: Elephant Corridors of India.* Delhi: Wildlife Trust of India, 2005.

Midgley, Mary. *Animals and Why They Matter.* London: Penguin, 1983.

Mikota, Susan. "Review of Tuberculosis in Captive Elephants and Implications for Wild Populations." *Gajah* 28 (2008): 8–18.

Miller, James. "Biodiversity Conservation and the Extinction of Experience." *Trends in Ecology and Evolution* 20, no. 8 (2005): 430–34.

Milton, Kay. *Loving Nature: Towards an Ecology of Emotion.* London: Routledge, 2002.

Mitman, Greg. "Pachyderm Personalities: The Media of Science, Politics and Conservation." In *Thinking with Animals: New Perspectives on Anthropomorphism,* edited by Gregg Mitman and Lorraine Daston, 175–95. New York: Columbia University Press, 2005.

———. *Reel Nature: America's Romance with Wildlife on Films.* Cambridge, Mass.: Harvard University Press, 1999.

Mol, Anne-Marie. *The Body Multiple: Ontology in Medical Practice.* Science and Cultural Theory. Durham, N.C.: Duke University Press, 2002.

———. "Ontological Politics: A Word and Some Questions." In *Actor-Network Theory and After,* edited by J. Law and J. Hassard, 74–89. Oxford and Keele: Blackwell and Sociological Review, 1999.

Mol, Anne-Marie, and John Law. "Regions, Networks and Fluids: Anemia and Social Topology." *Social Studies of Science* 24, no. 4 (1994): 641–71.

Monbiot, George. *Feral: Searching for Enchantment on the Frontiers of Rewilding.* London: Penguin, 2013.

Mora, C., D. P. Tittensor, S. Adl, A. G. B. Simpson, and B. Worm. "How Many Species Are There on Earth and in the Ocean?" *PLoS Biology* 9, no. 8 (2011).

Morris, Carol, and Matt Reed. "From Burgers to Biodiversity? The McDonaldization of On-Farm Nature Conservation in the UK." *Agriculture and Human Values* 24, no. 2 (2007): 207–18.

Morton, Timothy. *The Ecological Thought.* Cambridge, Mass.: Harvard University Press, 2010.

Murdoch, Jonathan. *Post-structuralist Geography: A Guide to Relational Space.* London: Sage, 2006.

Nagy, K., P. D. Johnson, and R. Malamud. *Trash Animals: How We Live with Nature's Filthy, Feral, Invasive, and Unwanted Species.* Minneapolis: University of Minnesota Press, 2013.

Nast, Heidi. "Critical Pet Studies?" *Antipode* 38, no. 5 (2006): 894–906.

Navarro, L. M., and H. M. Pereira. "Rewilding Abandoned Landscapes in Europe." *Ecosystems* 15, no. 6 (2012): 900–912.

Nelson, B., H. Hughes, R. Nash, and M. Warren. "*Leptidea reali* Reissinger 1989: A Butterfly New to Britain and Ireland." *Entomologist's Record* 113 (2001): 97–102.

Neves, Katja. "Cashing In on Cetourism: A Critical Ecological Engagement with Dominant E-Ngo Discourses on Whaling, Cetacean Conservation, and Whale Watching." *Antipode* 42, no. 3 (2010): 719–41.

Nicholls, Henry. *The Way of the Panda: The Curious History of China's Political Animal.* London: Profile, 2010.

Nordhaus, Ted, and Michael Shellenberger. *Break Through: From the Death of Environmentalism to the Politics of Possibility.* Boston: Houghton Mifflin, 2007.

Norris, Tony. "Report on the Corncrake." *British Birds* 38 (1945): 142–68.

———. "Report on the Distribution and Status of the Corncrake." *British Birds* 40 (1947): 226–44.

Ogden, Laura, Nik Heynen, Ulrich Oslender, Paige West, Karim-Aly Kassam, and Paul Robbins. "Global Assemblages, Resilience, and Earth Stewardship in the Anthropocene." *Frontiers in Ecology and the Environment* 11, no. 7 (2013): 341–47.

Olden, Julian. "Biotic Homogenization: A New Research Agenda for Conservation Biogeography." *Journal of Biogeography* 33, no. 12 (D2006): 2027–39.

Painlevé, Jean, Andy Masaki Bellows, Marina McDougall, and Brigitte Berg. *Science Is Fiction: The Films of Jean Painlevé.* Cambridge, Mass.: MIT Press, 2000.

Parmesan, Camille, and Gary Yohe. "A Globally Coherent Fingerprint of Climate Change Impacts across Natural Systems." *Nature* 421, no. 6918 (2003): 37–42.

Paterson, Mark. *Consumption and Everyday Life.* London: Taylor & Francis, 2005.

Phalan, B., M. Onial, A. Balmford, and R. E. Green. "Reconciling Food Production and Biodiversity Conservation: Land Sharing and Land Sparing Compared." *Science* 333, no. 6047 (2011): 1289–91.

Philo, Chris, and Chris Wilbert. *Animal Spaces, Beastly Places: New Geographies of Human–Animal Relations.* London: Routledge, 2000.

Pine, Joseph, and James H. Gilmore. *The Experience Economy: Work Is*

Theatre & Every Business a Stage. Boston: Harvard Business School Press, 1999.

Pink, Sarah. *Doing Visual Ethnography: Images, Media, and Representation in Research.* London: Sage, 2007.

Plumwood, Val. *Environmental Culture: The Ecological Crisis of Reason.* Environmental Philosophies Series. London: Routledge, 2002.

Pollan, Michael. *Second Nature: A Gardener's Education.* New York: Dell Publishing, 1991.

Popper, Karl. *The Open Society and Its Enemies.* 1947. London: Taylor & Francis, 2012.

Primack, Richard. *A Primer of Conservation Biology.* Sunderland, Mass.: Sinauer Associates, 1995.

Raffles, Hugh. *Insectopedia.* New York: Knopf Doubleday, 2010.

Rajan, Kaukik Sunder. *Lively Capital: Biotechnologies, Ethics, and Governance in Global Markets.* Durham, N.C.: Duke University Press, 2012.

Rancière, Jacques. *The Politics of Aesthetics: The Distribution of the Sensible.* London: Continuum, 2004.

Rechtbank's-Gravenhage. *Uitspraak in kort geding over de noodzaak tot bijvoeren van grote grazers (o.a. edelherten) in de Oostvaardersplassen in het licht van wettelijke zorgplichten.* LJN: AV4486, KG 06/171. March 13, 2006.

Reekie, Duncan. *Subversion: The Definitive History of Underground Cinema.* London: Wallflower Press, 2007.

Revkin, Andrew. "Emma Marris: In Defense of Everglades Pythons." *New York Times,* August 17, 2012.

Rheinberger, Hans-Jörg. *Toward a History of Epistemic Things: Synthesizing Proteins in the Test Tube.* Stanford, Calif.: Stanford University Press, 1997.

Ricciardi, Anthony, and Daniel Simberloff. "Assisted Colonization Is Not a Viable Conservation Strategy." *Trends in Ecology & Evolution* 24, no. 5 (2009): 248–53.

Ripple, William, and Robert Beschta. "Wolves and the Ecology of Fear: Can Predation Risk Structure Ecosystems?" *BioScience* 54, no. 8 (2004): 755–66.

Ritzer, George. *The McDonaldization Thesis.* London: Sage, 1988.

Roberts, Michael Symmons, and Paul Farley. *Edgelands.* New York: Random House, 2011.

Robertson, Morgan. "Measurement and Alienation: Making a World of Ecosystem Services." *Transactions of the Institute of British Geographers* 37, no. 3 (2012): 386–401.

———. "The Nature That Capital Can See: Science, State, and Market in the Commodification of Ecosystem Services." *Environment and Planning D: Society and Space* 24, no. 3 (2006): 367–87.

Rockstrom, Johan, Will Steffen, Kevin Noone, Asa Persson, F. Stuart Chapin, Eric F. Lambin, Timothy M. Lenton, et al. "A Safe Operating Space for Humanity." *Nature* 461, no. 7263 (2009): 472–75.

Rodriguez, J. P., A. B. Taber, P. Daszak, R. Sukumar, C. Valladares-Padua, S. Padua, L. F. Aguirre, et al. "Environment—Globalization of Conservation: A View from the South." *Science* 317, no. 5839 (2007): 755–56.

Rojas, M. "The Species Problem and Conservation: What Are We Protecting?" *Conservation Biology* 6, no. 2 (1992): 170–78.

Rose, Gillian. *Visual Methodologies: An Introduction to Researching with Visual Materials.* London: Sage, 2012.

Rose, Nikolas. *The Politics of Life Itself: Biomedicine, Power, and Subjectivity in the Twenty-First Century.* Princeton, N.J.: Princeton University Press, 2006.

Rosenzweig, Michael. "The Four Questions: What Does the Introduction of Exotic Species Do to Diversity?" *Evolutionary Ecology Research* 3, no. 3 (2001): 361–67.

———. "Reconciliation Ecology and the Future of Species Diversity." *Oryx* 37, no. 2 (2003): 194–205.

———. *Win-Win Ecology: How the Earth's Species Can Survive in the midst of Human Enterprise.* Oxford: Oxford University Press, 2003.

Roth, Wolff-Michael, and George Bowen. "Digitizing Lizards: The Topology of 'Vision' in Ecological Fieldwork." *Social Studies of Science* 29, no. 5 (1999): 719–64.

Rothfels, Nigel. "The Eyes of Elephants: Changing Perceptions." *Tidsskrift for Kulturforskning* 7, no. 3 (2008): 39–50.

Rothschild, B. M., and R. Laub. "Hyperdisease in the Late Pleistocene: Validation of an Early 20th Century Hypothesis." *Die Naturwissenschaften* 93, no. 11 (2006): 557–64.

RSPB. "Farms, Crofts & Corncrakes: A Guide to Habitat Management for Corncrakes." Sandy, U.K.: RSPB.

———. "The Future of Esas in Scotland." Sandy, U.K.: RSPB, 2003.

Rudel, T. K., L. Schneider, M. Uriarte, B. L. Turner Ii, R. DeFries, D. Lawrence, J. Geoghegan, et al. "Agricultural Intensification and Changes in Cultivated Areas, 1970–2005." *Proceedings of the National Academy of Sciences of the United States of America* 106, no. 49 (2009): 20675–80.

Russell, Catherine. *Experimental Ethnography.* Durham, N.C.: Duke University Press, 1999.

Rutherford, Stephanie. *Governing the Wild: Ecotours of Power.* Minneapolis: University of Minnesota Press, 2011.

Ryan, James. *Picturing Empire: Photography and the Visualization of the British Empire.* Chicago: University of Chicago Press, 1997.

Sachs, Jeffrey. *Common Wealth: Economics for a Crowded Planet.* New York: Penguin, 2008.

Santiapillai, C., P. Fernando, and M. Gunewardene. "A Strategy for the Conservation of the Asian Elephant in Sri Lanka." *Gajah* 25 (2006): 91–102.

Sax, D. F., and S. D. Gaines. "Species Diversity: From Global Decreases to Local Increases." *Trends in Ecology & Evolution* 18, no. 11 (2003): 561–66.

Schlaepfer, M. A., D. F. Sax, and J. D. Olden. "The Potential Conservation Value of Non-native Species." *El Valor de Conservación Potencial de Especies No Nativas* 25, no. 3 (2011): 428–37.

Scruton, Roger. *Animal Rights and Wrongs.* London: Metro, 2000.

Serpell, James. "Anthropomorphism and Anthropomorphic Selection: Beyond the 'Cute Response.'" *Society and Animals* 11, no. 1 (2003): 83–100.

Shapiro, Michael. *Cinematic Geopolitics.* New York: Routledge, 2008.

Shellenberger, Michael, and Ted Nordhaus, eds. *Love Your Monsters: Postenvironmentalism and the Anthropocene.* Oakland, Calif: Breakthrough Institute, 2011.

Shukin, Nicole. *Animal Capital: Rendering Life in Biopolitical Times.* Minneapolis: University of Minnesota Press, 2009.

Simpson, Kay. "'Doing Development': The Gap Year, Volunteer-Tourists and a Popular Practice of Development." *Journal of International Development* 16, no. 5 (2004): 681–92.

Sinclair, Ian. *Ghost Milk: Calling Time on the Grand Project.* London: Penguin, 2012.

Smit, R. *Oostvaardersplassen/Druk 1: Voorbij De Horizon Van Het Vertrouwde.* Driebergen, Netherlands: Staatsbosbeheer, 2010.

Smith, Richard. "World City Actor-Networks." *Progress in Human Geography* 27, no. 1 (2003): 25–44.

Sobchack, Vivian. *Carnal Thoughts: Embodiment and Moving Image Culture.* Berkeley: University of California Press, 2004.

Song, Hoon. *Pigeon Trouble: Bestiary Biopolitics in a Deindustrialized America.* Philadelphia; University of Pennsylvania Press, 2011.

Soule, Michael. "The 'New Conservation.'" *Conservation Biology* 27, no. 5 (2013): 895–97.

———. "What Is Conservation Biology." *Bioscience* 35, no. 11 (1985): 727–34.

Soule, Michael, and Gary Lease. *Reinventing Nature? Responses to Postmodern Deconstruction.* Washington, D.C.: Island Press, 1995.

Soutullo, Alvaro. "Extent of the Global Network of Terrestrial Protected Areas." *Conservation Biology* 24, no. 2 (2010): 362–63.

Srinivasan, Krithika. "The Biopolitics of Animal Being and Welfare: Dog Control and Care in the UK and India." *Transactions of the Institute of British Geographers* 38, no. 1 (2013): 106–19.

Sri Lanka Department for Wildlife Conservation. *National Policy for the Conservation and Management of Wild Elephants in Sri Lanka.* Colombo, Sri Lanka: Department of Wildlife Conservation, 2007.

Star, Susan Leigh, and Jim Griesemer. "Institutional Ecology, Translations and Boundary Objects: Amateurs and Professionals in Berkeley's Museum of Vertebrate Zoology, 1907–39." *Social Studies of Science* 19, no. 3 (1989): 387–420.

Steffen, Will, Paul Crutzen, and John McNeill. "The Anthropocene: Are Humans Now Overwhelming the Great Forces of Nature?" *AMBIO: A Journal of the Human Environment* 36, no. 8 (2007): 614–21.

Steffen, Will, Jacques Grinevald, Paul Crutzen, and John McNeill. "The Anthropocene: Conceptual and Historical Perspectives." *Philosophical Transactions of the Royal Society A: Mathematical, Physical and Engineering Sciences* 369, no. 1938 (2011): 842–67.

Stengers, Isabelle. *Cosmopolitics.* Posthumanities. Minneapolis: University of Minnesota Press, 2010.

———. *Cosmopolitics: II.* Minneapolis: University of Minnesota Press, 2011.

Stewart, Kathleen. *Ordinary Affects.* Durham, N.C.: Duke University Press, 2007.

Stone, Phillip. "Dark Tourism Scholarship: A Critical Review." *International Journal of Culture, Tourism, and Hospitality Research* 7, no. 3 (2013): 307–18.

Stork, Nigel. "Measuring Global Biodiverity and Its Decline." In *Biodiversity II: Understanding and Protecting Our Biological Resources,* edited by Marjorie Reaka-Kudla, Don E. Wilson, and Edward O. Wilson, 41–68. Washington, D.C.: Joseph Henry Press, 1997.

Stowe, Tim, and Andrew Hudson. "Corncrake Studies in the Western Isles." *RSPB Conservation Review* 2 (1988): 38–42.

———. "Radio-Telemetry Studies of Corncrake in Great Britain." *Die Vogelwelte* 112, nos. 10–16 (1991).

Stubbs, Alan. *British Hoverflies.* Reading, U.K.: British Entomological & Natural History Society, 1996.

Sukumar, Raman. *The Living Elephants: Evolutionary Ecology, Behavior, and Conservation.* New York: Oxford University Press, 2003.

Sullivan, Sian. "Banking Nature? The Spectacular Financialisation of Environmental Conservation." *Antipode* 45, no. 1 (2013): 198–217.

Sutherland, William. "Conservation Biology: Openness in Management." *Nature* 418, no. 6900 (2002): 834–35.

Svenning, Jens-Christian. "A Review of Natural Vegetation Openness in North-western Europe." *Biological Conservation* 104, no. 2 (2002): 133–48.

Swyngedouw, Erik. "Apocalypse Forever? Post-political Populism and the Spectre of Climate Change." *Theory, Culture & Society* 27, nos. 2–3 (2010): 213–32.

Takacs, David. *The Idea of Biodiversity: Philosophies of Paradise.* Baltimore, Md.: Johns Hopkins University Press, 1996.

Thompson, Charis. *Making Parents: The Ontological Choreography of Reproductive Technologies.* Cambridge, Mass.: MIT Press, 2005.

———. "When Elephants Stand for Competing Philosophies of Nature: Amboseli National Park, Kenya." In *Complexities: Social Studies of Knowledge Practices,* edited by John Law and Anne-Marie Mol, 166–90. Durham, N.C.: Duke University Press, 2002.

Thrift, Nigel. "Intensities of Feeling: Towards a Spatial Politics of Affect." *Geografiska Annaler: Series B, Human Geography* 86, no. 1 (2004): 57–78.

———. *Non-representational Theory: Space, Politics, Affect.* London: Routledge, 2007.

Tomazos, Kostas, and Richard Butler. "The Volunteer Tourist as 'Hero.'" *Current Issues in Tourism* 13, no. 4 (2010): 363–80.

Toropova, Caitlyn, Imèn Meliane, Dan Laffoley, Elizabeth Matthews, and Mark Spalding. *Global Ocean Protection: Present Status and Future Possibilities.* Gland, Switzerland: IUCN, 2010.

Totaro, Donato, "Deleuzian Film Analysis: The Skin of Film." *Offscreen* 6, no. 6 (2002).

Tsing, Anna. *Friction: An Ethnography of Global Connection.* Princeton, N.J.: Princeton University Press, 2005.

———. "Unruly Edges: Mushrooms as Companion Species." Environmental Humanities 1 (2012): 141–54.

Tuan, Yi-fu. *Topophilia: A Study of Environmental Perception, Attitudes, and Values.* Englewood Cliffs, N.J.: Prentice-Hall, 1974.

Turnbaugh, Peter J., Ruth E. Ley, Micah Hamady, Claire M. Fraser-Liggett, Rob Knight, and Jeffrey I. Gordon. "The Human Microbiome Project." *Nature* 449, no. 7164 (2007): 804–10.

Turnhout, Esther. *Ecological Indicators in Dutch Nature Conservation: Science and Policy Intertwined in the Classification and Evaluation of Nature.* Amsterdam: Aksant, 2003.

Tyler, G. A., and R. E. Green. "The Incidence of Nocturnal Song by Male Corncrakes *Crex crex* Is Reduced during Pairing." *Bird Study* 43 (1996): 214–19.

Tyler, G. A., R. E. Green, and C. Casey. "Survival and Behaviour of Corncrake *Crex crex* Chicks during the Mowing of Agricultural Grassland." *Bird Study* 45 (1998): 35–50.

Tyler, Tom. "If Horses Had Hands." *Society and Animals* 11, no. 3 (2003): 267–81.

UKBP. "Corncrake Species Action Plan." London: HMSO, 2004.

Urban Task Force. *Towards an Urban Renaissance.* London: Taylor & Francis, 2003.

Vera, Frans. *Grazing Ecology and Forest History.* Wallingford, U.K.: CABI, 2000.

Vera, F., F. Buissink, and J. Weidema. *Wilderness in Europe: What Really Goes on between the Trees and the Beasts?* Baarn, Netherlands: Tirion, 2007.

Vidya, T. N., R. Sukumar, and D. J. Melnick. "Range-wide Mtdna Phylogeography Yields Insights into the Origins of Asian Elephants." *Proceedings: Biological Sciences: The Royal Society* 276, no. 1658 (2009): 893–902.

von Uexkull, Jakob and J. D. O'Neil. *A Foray into the Worlds of Animals and Humans: With a Theory of Meaning.* Minneapolis: University of Minnesota Press, 2010.

Vulink, J. "Hungry Herds: The Management of Temperate Lowland Wetlands by Grazing." PhD diss., Rijksuniversiteit Groningen, Netherlands, 2001.

Walker, Jeremy, and Melinda Cooper. "Genealogies of Resilience: From Systems Ecology to the Political Economy of Crisis Adaptation." *Security Dialogue* 42, no. 2 (2011): 143–60.

Wapner, Paul. *Living through the End of Nature: The Future of American Environmentalism.* Cambridge, Mass.: MIT Press, 2010.

Warren, Charles. "Perspectives on the 'Alien' versus 'Native' Species Debate: A Critique of Concepts, Language and Practice." *Progress in Human Geography* 31, no. 4 (2007): 427–46.

Waterton, Clare. "From Field to Fantasy: Classifying Nature, Constructing Europe." *Social Studies of Science* 32, no. 2 (April 2002): 177–204.

Waterton, Clare, Rebecca Ellis, and Brian Wynne. *Barcoding Nature: Shifting Cultures of Taxonomy in an Age of Biodiversity Loss.* London: Taylor & Francis, 2013.

Wearing, Stephen. *Volunteer Tourism: Experiences That Make a Difference.* Wallingford, U.K.: CABI, 2001.

Weber, Max. "The Sociology of Charismatic Authority." In *From Max Weber,* edited by H. Gerth and C. Mills, 245–52. New York: Oxford University Press, 1958.

Weber, Max, and S. N. Eisenstadt. *Max Weber on Charisma and Institution Building: Selected Papers.* Chicago: University of Chicago Press, 1968.

Weinstein, Paul. "Insects in Psychiatry." *Cultural Entomology Digest* 2 (1994).

Weisman, Alan. *The World without Us.* London: Ebury Publishing, 2012.

West, Paige. *Conservation Is Our Government Now: The Politics of Ecology in Papua New Guinea.* Durham, N.C.: Duke University Press, 2006.

West, Paige, and James Carrier. "Ecotourism and Authenticity: Getting Away from It All?" *Current Anthropology* 45, no. 4 (2004): 483–98.

Whatmore, Sarah. *Hybrid Geographies: Natures, Cultures, Spaces.* London: Sage, 2002.

———. "Living Cities: Making Space for Urban Nature." *Soundings: Journal of Politics and Culture* 22 (2002): 137–50.

Whatmore, Sarah, and Bruce Braun. *Political Matter: Technoscience, Democracy and Public Life.* Oxford: Oxford University Press, 2010.

Whatmore, Sarah, and Lorraine Thorne. "Elephants on the Move: Spatial Formations of Wildlife Exchange." *Environment and Planning D: Society and Space* 18, no. 2 (2000): 185–203.

———. "Wild(er)ness: Reconfiguring the Geographies of Wildlife." *Transactions of the Institute of British Geographers* 23, no. 4 (1998): 435–54.

Whittaker, R., M. Araujo, P. Jepson, R. Ladle, J. Watson, and K. Willis. "Conservation Biogeography: Assessment and Prospect." *Diversity and Distributions* 11, no. 1 (2005): 3–23.

Williams, G., R. Green, C. Casey, B. Deceuninck, and T. Stowe. "Halting Declines in Globally Threatened Species: The Case of the Corncrake." *RSPB Conservation Review* 11 (1997): 22–31.

Willis, K. J., and H. J. B. Birks. "What Is Natural? The Need for a Long-Term Perspective in Biodiversity Conservation." *Science* 314, no. 5803 (2006): 1261–65.

Wilson, Alexander. *The Culture of Nature: North American Landscape from Disney to the Exxon Valdez.* Cambridge: Blackwell, 1992.

Wilson, Edward. *Biodiversity.* Washington, D.C.: National Academy Press, 1988.

———. *Biophilia.* Cambridge, Mass.: Harvard University Press, 1984.

———. *The Diversity of Life.* Questions of Science. Cambridge, Mass.: Belknap Press of Harvard University Press, 1992.

———. *Naturalist.* Washington, D.C.: Island Press, 1994.

Wolch, Jennifer. "Zoopolis." In *Animal Geographies: Place, Politics and Identity in the Nature–Culture Borderlands,* edited by Jennifer Wolch and Jody Emel, 119–38. London: Verso Books, 1998.

Wolfe, Cary. *Before the Law: Humans and Other Animals in a Biopolitical Frame.* Chicago: University of Chicago Press, 2012.

———. *Zoontologies: The Question of the Animal.* Minneapolis: University of Minnesota Press, 2003.

Woodard, Ben. *On an Ungrounded Earth: Towards a New Geophilosophy.* Brooklyn, N.Y.: Punctum Books, 2013.

Worster, Donald. "Nature and the Disorder of History." In *Reinventing Nature,* edited by Michael E. Soule and Gary Lease, 65–85. Washington, D.C.: Island Press, 1995.

Youatt, Rafi. "Counting Species: Biopower and the Global Biodiversity Census." *Environmental Values* 17, no. 3 (2008): 393–417.

Yusoff, Kathryn. "Aesthetics of Loss: Biodiversity, Banal Violence and Biotic Subjects." *Transactions of the Institute of British Geographers* 37, no. 4 (2012): 578–92.

———. "Biopolitical Economies and the Political Aesthetics of Climate Change." *Theory, Culture and Society* 27, no. 2 (2010): 73–99.

———. "Geologic Life: Prehistory, Climate, Futures in the Anthropocene." *Environment and Planning D: Society and Space* 31, no. 5 (2013): 779–95.

Zalasiewicz, Jan, Mark Williams, Richard Fortey, Alan Smith, Tiffany L. Barry, Angela L. Coe, Paul R. Bown, et al. "Stratigraphy of the Anthropocene." *Philosophical Transactions of the Royal Society A: Mathematical, Physical and Engineering Sciences* 369, no. 1938 (2011): 1036–55.

Zalasiewicz, Jan, Mark Williams, Alan Haywood, and Michael Ellis. "The Anthropocene: A New Epoch of Geological Time?" *Philosophical Transactions of the Royal Society A: Mathematical, Physical and Engineering Sciences* 369, no. 1938 (2011): 835–41.

Zimmerer, Karl. "The Reworking of Conservation Geographies: Nonequilibrium Landscapes and Nature–Society Hybrids." *Annals of the Association of American Geographers* 90, no. 2 (2000): 356–69.

Zimov, Sergei. "Pleistocene Park: Return of the Mammoth's Ecosystem." *Science* 308, no. 5723 (2005): 796–98.

Žižek, Slavoj. *In Defense of Lost Causes.* London: Verso Books, 2008.

———. *Living in the End Times.* London: Verso Books, 2010.

INDEX

abandonment, ecologies of, 21, 166–67. *See also* urban wildlife/ecologies

abstraction, affective dimensions of, 53–54, 207n48

actor–network theory (ANT), 26, 171

adaptation: to anticipated "climate envelopes," 171; dynamic charm as sense of "response-ability" and, 226n2; equilibrium ecology and inability to learn from ecological, 95–96; fluid topology and, 167, 174–75. *See also* climate change

adventure, voluntourism and, 149, 151

aesthetic charisma, 40, 44–50; defined, 44; divergent popular responses and, 46–50; negative, 48–49. *See also* imagery in nature conservation, role of moving

aesthetics, political, 122

affect: concept of, 44–45; relational account of, 45–51. *See also* affective logic(s)

"affection-images," 125

affective logic(s), 9; affect, film, and biopolitics, 121–23; of awe, 130–33, 137; characterizing wildlife film, 17; of curiosity, 49, 131, 133–36, 137–38, 182; defined, 39, 122; epiphany, 50, 51–52; of farming and hunting, 50; framing interspecies encounters in conservation, 9–12, 13, 148–52; image wars relating to OVP and, 114; impact on conservation, 46; jouissance, 50, 52–54, 207n44; of lab and field, different,

53–54; "learning to be affected" by environment, 5–6, 9–12, 35–38, 43; micropolitics of elephant imagery and, 119–38, 217n25, 218n28; passions powering conservation, 38; reason or rationality, 45; of sentimentality, 124–27; of the sublime, 131, 132, 150, 151; of sympathy, 127–30, 137. *See also* charisma, nonhuman

Africa: persistence of exoticized imaginations of, 148; as wilderness, politics of framing, 219n50

Africa (BBC series), 133

African elephants, 226n10; *Echo and Other Elephants* (documentary) on, 127–28, 130; as ecological agents, 187; exposed to poaching and culls, depression and rage among, 27; as focal species for rewilding/reintroduction, 187–88, 215n32

Agamben, Giorgio, 200n60, 205n9; on "anthropological machine," 13–14

agency, nonhuman, 39, 54, 58, 103, 104, 109, 141, 155, 180, 183, 187, 203n52; Asian elephant conservation and, 25–28; tendency to downplay, in voluntourism, 147–48

agriculture: affective logics of farming, 50; corncrake action plan implementation and, 90–93; crofting, 16, 77, 78, 84, 87–88, 91–94, 210n1; framing corncrake as casualty of intensification of, 86–88; in India,

Jamie Lorimer is associate professor of human geography at the University of Oxford and a tutorial fellow at Hertford College.